An Introduction to Causal Inference

Judea Pearl

Copyright © 2015

All rights reserved.

ISBN: 1507894295
ISBN-13: 978-1507894293
DOI: 10.2202/1557-4679.1203

The publisher does not warrant that the information contained in the book is complete and correct and shall not be liable whatsoever for any damages incurred as a result of its use. The views expressed herein are not the views of the publisher. Time sensitive content is not meant to be taken as valid after the date of publication. Any fictional depictions are not meant to represent any real person dead, or living. It is not the intent to malign any religion, ethnicity, or organization. The appearance of external hyperlinks or text does not constitute endorsement by any organization of the publisher, or the information, products or services contained herein. Unless otherwise specified, the organizations, authors, and sources do not exercise any editorial control over the information you may find herein. No part of this book may be used or reproduced in any manner whatsoever without written permission except in the case of brief quotations embodied in critical articles or reviews.

CONTENTS

1	Abstract	1
2	Introduction	3
3	From Association to Causation	5
4	Structural Models, Diagrams, Causal Effects, and Counterfactuals	9
5	Methodological Principles of Causal Inference	35
6	The Potential Outcome Framework	41
7	Counterfactuals at Work	49
8	Conclusions	67
9	Footnotes	69

Abstract

This paper summarizes recent advances in causal inference and underscores the paradigmatic shifts that must be undertaken in moving from traditional statistical analysis to causal analysis of multivariate data. Special emphasis is placed on the assumptions that underlie all causal inferences, the languages used in formulating those assumptions, the conditional nature of all causal and counterfactual claims, and the methods that have been developed for the assessment of such claims. These advances are illustrated using a general theory of causation based on the Structural Causal Model (SCM) described in Pearl (2000a), which subsumes and unifies other approaches to causation, and provides a coherent mathematical foundation for the analysis of causes and counterfactuals. In particular, the paper surveys the development of mathematical tools for inferring (from a combination of data and assumptions) answers to three types of causal queries: those about (1) the effects of potential interventions, (2) probabilities of counterfactuals, and (3) direct and indirect effects (also known as "mediation"). Finally, the paper defines the formal and conceptual relationships between the structural and potential-outcome frameworks and presents tools for a symbiotic analysis that uses the strong features of both. The tools are demonstrated in the analyses of mediation, causes of effects, and probabilities of causation.

Judea Pearl

1. Introduction

Most studies in the health, social and behavioral sciences aim to answer *causal* rather than associative – questions. Such questions require some knowledge of the data-generating process, and cannot be computed from the data alone, nor from the distributions that govern the data. Remarkably, although much of the conceptual framework and algorithmic tools needed for tackling such problems are now well established, they are not known to many of the researchers who could put them into practical use. Solving causal problems systematically requires certain extensions in the standard mathematical language of statistics, and these extensions are not typically emphasized in the mainstream literature. As a result, many statistical researchers have not yet benefited from causal inference results in (i) counterfactual analysis, (ii) nonparametric structural equations, (iii) graphical models, and (iv) the symbiosis between counterfactual and graphical methods. This survey aims at making these contemporary advances more accessible by providing a gentle introduction to causal inference for a more in-depth treatment and its methodological principles (see (Pearl, 2000a, 2009a,b)).

In Section 2, we discuss coping with untested assumptions and new mathematical notation which is required to move from associational to causal statistics. Section 3.1 introduces the fundamentals of the structural theory of causation and uses these modeling fundamentals to represent interventions and develop mathematical tools for estimating causal effects (Section 3.3) and counterfactual quantities (Section 3.4). Section 4 outlines a general methodology to guide problems of causal inference: Define, Assume, Identify and Estimate, with each step benefiting from the tools developed in Section 3.

Section 5 relates these tools to those used in the potential-outcome framework, and offers a formal mapping between the two frameworks and a symbiosis (Section 5.3) that exploits the best features of both. Finally, the benefit of this symbiosis is demonstrated in Section 6, in which the structure-based logic of counterfactuals is harnessed to estimate causal quantities that cannot be defined within the paradigm of controlled randomized experiments. These include direct and indirect effects, the effect of treatment on the treated, and questions of attribution, i.e., whether one event can be deemed "responsible" for another.

2. From Association to Causation
2.1. Understanding the distinction and its implications

The aim of standard statistical analysis is to assess parameters of a distribution from samples drawn of that distribution. With the help of such parameters, associations among variables can be inferred, which permits the researcher to estimate probabilities of past and future events and update those probabilities in light of new information. These tasks are managed well by standard statistical analysis so long as experimental conditions remain the same. Causal analysis goes one step further; its aim is to infer probabilities under conditions that are *changing*, for example, changes induced by treatments or external interventions.

This distinction implies that causal and associational concepts do not mix; there is nothing in a distribution function to tell us how that distribution would differ if external conditions were to change—say from observational to experimental setup—because the laws of probability theory do not dictate how one property of a distribution ought to change when another property is modified. This information must be provided by causal assumptions which identify relationships that remain invariant when external conditions change.

A useful demarcation line between associational and causal concepts crisp and easy to apply, can be formulated as follows. An associational concept is any relationship that can be defined in terms of a joint distribution of observed variables, and a causal concept is any relationship that cannot be defined from the distribution alone. Examples of associational concepts are: correlation, regression, dependence, conditional independence, likelihood, collapsibility, propensity score, risk ratio, odds ratio, marginalization, conditionalization, "controlling for," and many more. Examples of causal concepts are: randomization, influence, effect, confounding, "holding constant," disturbance, error terms, structural coefficients, spurious correlation, faithfulness/stability, instrumental variables, intervention, explanation, and attribution. The former can, while the latter cannot be defined in term of distribution functions.

This demarcation line is extremely useful in tracing the assumptions that are needed for substantiating various types of scientific claims. Every claim invoking causal concepts must rely on some premises that invoke such concepts; it cannot be inferred from, or even defined in terms statistical associations alone.

This distinction further implies that causal relations cannot be expressed in the language of probability and, hence, that any mathematical approach to causal analysis must acquire new notation – probability calculus is insufficient. To illustrate, the syntax of probability calculus does not permit us to express the simple fact that "symptoms do not cause diseases," let alone draw mathematical conclusions from such facts. All we can say is that two events are dependent—meaning that if we find one, we can expect to encounter the other, but we cannot distinguish statistical dependence, quantified by the conditional probability $P(disease|symptom)$ from causal dependence, for which we have no expression in standard probability calculus.

2.2. Untested assumptions and new notation

The preceding two requirements: (1) to commence causal analysis with untested,1 theoretically or judgmentally based assumptions, and (2) to extend the syntax of probability calculus, constitute the two primary barriers to the acceptance of causal analysis among professionals with traditional training in statistics.

Associational assumptions, even untested, are testable in principle, given sufficiently large sample and sufficiently fine measurements. Causal assumptions, in contrast, cannot be verified even in principle, unless one resorts to experimental control. This difference stands out in Bayesian analysis. Though the priors that Bayesians commonly assign to statistical parameters are untested quantities, the sensitivity to these priors tends to diminish with increasing sample size. In contrast, sensitivity to prior causal assumptions, say that treatment does not change gender, remains substantial regardless of sample size.

This makes it doubly important that the notation we use for expressing causal assumptions be cognitively meaningful and unambiguous so that one can clearly judge the plausibility or inevitability of the assumptions articulated. Statisticians can no longer ignore the mental representation in which scientists store experiential knowledge, since it is this representation, and the language used to access it that determine the reliability of the judgments upon which the analysis so crucially depends.

Those versed in the potential-outcome notation (Neyman, 1923, Rubin, 1974, Holland, 1988), can recognize causal expressions through the subscripts that are attached to counterfactual events and variables, e.g. $Y_x(u)$ or Z_{xy}. (Some authors use parenthetical expressions, e.g. $Y(0)$, $Y(1)$, $Y(x, u)$ or $Z(x, y)$.) The expression $Y_x(u)$, for example, stands for the value that outcome Y would take in individual u, had treatment X been at level x. If u is chosen at random, Y_x is a random variable, and one can talk about the probability that Y_x would attain a value y in the population, written $P(Y_x = y)$ (see Section 5 for semantics). Alternatively, Pearl (1995) used expressions of the form $P(Y = y | set(X = x))$ or $P(Y = y | do(X = x))$ to denote the probability (or frequency) that event $(Y = y)$ would occur if treatment condition $X = x$ were enforced uniformly over the population.2 Still a third notation that distinguishes causal expressions is provided by graphical models, where the arrows convey causal directionality.

However, few have taken seriously the textbook requirement that any introduction of new notation must entail a systematic definition of the syntax and semantics that governs the notation. Moreover, in the bulk of the statistical literature before 2000, causal claims rarely appear in the mathematics. They surface only in the verbal interpretation that investigators occasionally attach to certain associations, and in the verbal description with which investigators justify assumptions. For example, the assumption that a covariate not be affected by a treatment, a necessary assumption for the control of confounding (Cox, 1958, p. 48), is expressed in plain English, not in a mathematical expression.

The next section provides a conceptualization that overcomes these mental barriers by offering a friendly mathematical machinery for cause-effect analysis and a formal foundation for counterfactual analysis.

3. Structural Models, Diagrams, Causal Effects, and Counterfactuals

Any conception of causation worthy of the title "theory" must be able to (1) represent causal questions in some mathematical language, (2) provide a precise language for communicating assumptions under which the questions need to be answered, (3) provide a systematic way of answering at least some of these questions and labeling others "unanswerable," and (4) provide a method of determining what assumptions or new measurements would be needed to answer the "unanswerable" questions.

A "general theory" should do more. In addition to embracing *all* questions judged to have causal character, a general theory must also *subsume* any other theory or method that scientists have found useful in exploring the various aspects of causation. In other words, any alternative theory needs to evolve as a special case of the "general theory" when restrictions are imposed on either the model, the type of assumptions admitted, or the language in which those assumptions are cast.

The structural theory that we use in this survey satisfies the criteria above. It is based on the Structural Causal Model (SCM) developed in (Pearl, 1995, 2000a) which combines features of the structural equation models (SEM) used in economics and social science (Goldberger, 1973, Duncan, 1975), the potential-outcome framework of Neyman (1923) and Rubin (1974), and the graphical models developed for probabilistic reasoning and causal analysis (Pearl, 1988, Lauritzen, 1996, Spirtes, Glymour, and Scheines, 2000, Pearl, 2000a).

Although the basic elements of SCM were introduced in the mid 1990's (Pearl, 1995), and have been adapted widely by epidemiologists (Greenland, Pearl, and Robins, 1999, Glymour and Greenland, 2008), statisticians (Cox and Wermuth, 2004, Lauritzen, 2001), and social scientists (Morgan and Winship, 2007), its potentials as a comprehensive theory of causation are yet to be fully utilized. Its ramifications thus far include:

The unification of the graphical, potential outcome, structural equations, decision analytical (Dawid, 2002), interventional (Woodward, 2003), sufficient component (Rothman, 1976) and probabilistic (Suppes, 1970) approaches to causation; with each approach viewed as a restricted version of the SCM.

The definition, axiomatization and algorithmization of counterfactuals and joint probabilities of counterfactuals

Reducing the evaluation of "effects of causes," "mediated effects," and "causes of effects" to an algorithmic level of analysis.

Solidifying the mathematical foundations of the potential-outcome model, and formulating the counterfactual foundations of structural equation models.

Demystifying enigmatic notions such as "confounding," "mediation," "ignorability," "comparability," "exchangeability (of populations)," "superexogeneity" and others within a single and familiar conceptual framework.

Weeding out myths and misconceptions from outdated traditions (Meek and Glymour, 1994, Greenland et al., 1999, Cole and Hernán, 2002, Arah, 2008, Shrier, 2009, Pearl, 2009c).

This section provides a gentle introduction to the structural framework and uses it to present the main advances in causal inference that have emerged in the past two decades.

3.1. A brief introduction to structural equation models

How can one express mathematically the common understanding that symptoms do not cause diseases? The earliest attempt to formulate such relationship mathematically was made in the 1920's by the geneticist Sewall Wright (1921). Wright used a combination of equations and graphs to communicate causal relationships. For example, if X stands for a disease variable and Y stands for a certain symptom of the disease, Wright would write a linear equation:[3]

(1) $y = \beta x + u_Y$

where x stands for the level (or severity) of the disease, y stands for the level (or severity) of the symptom, and u_Y stands for all factors, other than the disease in question, that could possibly affect Y when X is held constant. In interpreting this equation one should think of a physical process whereby Nature *examines* the values of x and u and, accordingly, *assigns* variable Y the value $y = \beta x + u_Y$. Similarly, to "explain" the occurrence of disease X, one could write $x = u_X$, where U_X stands for all factors affecting X.

Equation (1) still does not properly express the causal relationship implied by this assignment process, because algebraic equations are symmetrical objects; if we re-write (1) as

(2) $x = (y - u_Y)/\beta$

it might be misinterpreted to mean that the symptom influences the disease. To express the directionality of the underlying process, Wright augmented the equation with a diagram, later called "path diagram," in which arrows are drawn from (perceived) causes to their (perceived) effects, and more importantly, the absence of an arrow makes the empirical claim that Nature assigns values to one variable irrespective of another. In Fig. 1, for example, the absence of arrow from Y to X represents the claim that symptom Y is not among the factors U_X which affect disease X. Thus, in our example, the complete model of a symptom and a disease would be written as in Fig. 1: The diagram encodes the possible existence of (direct) causal influence of X on Y, and the absence of causal influence of Y on X, while the equations encode the quantitative relationships among the variables involved, to be determined from the data. The parameter β in the equation is called a "path coefficient" and it quantifies the (direct) causal effect of X on Y; given the numerical values of β and U_Y, the equation claims that, a unit increase for X would result in β units increase of Y regardless of the values taken by other variables in the model, and regardless of whether the increase in X originates from external or internal influences.

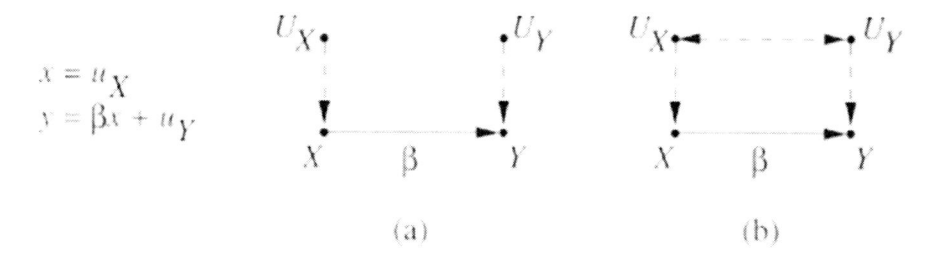

Figure 1: A simple structural equation model, and its associated diagrams. Unobserved exogenous variables are connected by dashed arrows.

The variables U_X and U_Y are called "exogenous;" they represent observed or unobserved background factors that the modeler decides to keep unexplained, that is, factors that influence but are not influenced by the other variables (called "endogenous") in the model. Unobserved exogenous variables are sometimes called "disturbances" or "errors", they represent factors omitted from the model but judged to be relevant for explaining the behavior of variables in the model. Variable U_X, for example, represents factors that contribute to the disease X, which may or may not be correlated with U_Y (the factors that influence the symptom Y). Thus, background factors in structural equations differ fundamentally from residual terms in regression equations. The latters are artifacts of analysis which, by definition, are uncorrelated with the regressors. The formers are part of physical reality (e.g., genetic factors, socio-economic conditions) which are responsible for variations observed in the data; they are treated as any other variable, though we often cannot measure their values precisely and must resign to merely acknowledging their existence and assessing qualitatively how they relate to other variables in the system.

If correlation is presumed possible, it is customary to connect the two variables, U_Y and U_X, by a dashed double arrow, as shown in Fig. 1(b).

In reading path diagrams, it is common to use kinship relations such as parent, child, ancestor, and descendent, the interpretation of which is usually self evident. For example, an arrow $X \to Y$ designates X as a parent of Y and Y as a child of X. A "path" is any consecutive sequence of edges, solid or dashed. For example, there are two paths between X and Y in Fig. 1(b), one consisting of the direct arrow $X \to Y$ while the other tracing the nodes X, U_X, U_Y and Y.

Wright's major contribution to causal analysis, aside from introducing the language of path diagrams, has been the development of graphical rules for writing down the covariance of any pair of observed variables in terms of path coefficients and of covariances among the error terms. In our simple example, one can immediately write the relations

(3) $Cov(X, Y) = \beta$

for Fig. 1(a), and

(4) $Cov(X, Y) = \beta + Cov(U_Y, U_X)$

for Fig. 1(b) (These can be derived of course from the equations, but, for large models, algebraic methods tend to obscure the origin of the derived quantities). Under certain conditions, (e.g. if $Cov(U_Y, U_X) = 0$), such relationships may allow one to solve for the path coefficients in term of observed covariance terms only, and this amounts to inferring the magnitude of (direct) causal effects from observed, nonexperimental associations, assuming of course that one is prepared to defend the causal assumptions encoded in the diagram.

It is important to note that, in path diagrams, causal assumptions are encoded not in the links but, rather, in the missing links. An arrow merely indicates the possibility of causal connection, the strength of which remains to be determined (from data); a missing arrow represents a claim of zero influence, while a missing double arrow represents a claim of zero covariance. In Fig. 1(a), for example, the assumptions that permits us to identify the direct effect β are encoded by the missing double arrow between U_X and U_Y, indicating $Cov(U_Y, U_X)=0$, together with the missing arrow from Y to X. Had any of these two links been added to the diagram, we would not have been able to identify the direct effect β. Such additions would amount to relaxing the assumption $Cov(U_Y, U_X) = 0$, or the assumption that Y does not effect X, respectively. Note also that both assumptions are causal, not associational, since none can be determined from the joint density of the observed variables, X and Y; the association between the unobserved terms, U_Y and U_X, can only be uncovered in an experimental setting; or (in more intricate models, as in Fig. 5) from other causal assumptions.

Although each causal assumption in isolation cannot be tested, the sum total of all causal assumptions in a model often has testable implications. The chain model of Fig. 2(a), for example, encodes seven causal assumptions, each corresponding to a missing arrow or a missing double-arrow between a pair of variables. None of those assumptions is testable in isolation, yet the totality of all those assumptions implies that Z is unassociated with Y in every stratum of X. Such testable implications can be read off the diagrams using a graphical criterion known as *d-separation* (Pearl, 1988).

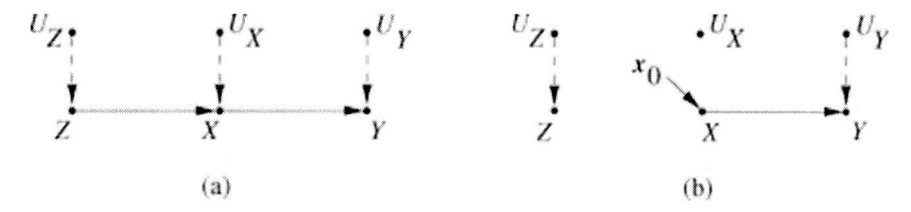

Figure 2: (a) The diagram associated with the structural model of Eq. (5). (b) The diagram associated with the modified model of Eq. (6), representing the intervention $do(X = x_0)$.

Definition 1 (d-separation) *A set S of nodes is said to block a path p if either (i) p contains at least one arrow-emitting node that is in S, or (ii) p contains at least one collision node that is outside S and has no descendant in S. If S blocks all paths from X to Y, it is said to "d-separate X and Y," and then, X and Y are independent given S, written $X \perp\!\!\!\perp Y | S$.*

To illustrate, the path $U_Z \rightarrow Z \rightarrow X \rightarrow Y$ is blocked by $S = \{Z\}$ and by $S = \{X\}$, since each emits an arrow along that path. Consequently we can infer that the conditional independencies $U_z \perp\!\!\!\perp Y|Z$ and $U_z \perp\!\!\!\perp Y|X$ will be satisfied in any probability function that this model can generate, regardless of how we parametrize the arrows. Likewise, the path $U_z \to Z \to X \leftarrow U_x$ is blocked by the null set $\{\emptyset\}$ but is not blocked by $S = \{Y\}$, since Y is a descendant of the collision node X. Consequently, the marginal independence $U_z \perp\!\!\!\perp U_x$ will hold in the distribution, but $U_z \perp\!\!\!\perp U_x|Y$ may or may not hold. This special handling of collision nodes (or colliders, e.g., $Z \to X \leftarrow U_x$) reflects a general phenomenon known as *Berkson's paradox* (Berkson, 1946), whereby observations on a common consequence of two independent causes render those causes dependent. For example, the outcomes of two independent coins are rendered dependent by the testimony that at least one of them is a tail.

The conditional independencies entailed by *d*-separation constitute the main opening through which the assumptions embodied in structural equation models can confront the scrutiny of nonexperimental data. In other words, almost all statistical tests capable of invalidating the model are entailed by those implications.[4]

3.2. From linear to nonparametric models and graphs

Structural equation modeling (SEM) has been the main vehicle for effect analysis in economics and the behavioral and social sciences (Goldberger, 1972, Duncan, 1975, Bollen, 1989). However, the bulk of SEM methodology was developed for linear analysis and, until recently, no comparable methodology has been devised to extend its capabilities to models involving dichotomous variables or nonlinear dependencies. A central requirement for any such extension is to detach the notion of "effect" from its algebraic representation as a coefficient in an equation, and redefine "effect" as a general capacity to transmit *changes* among variables. Such an extension, based on simulating hypothetical interventions in the model, was proposed in (Haavelmo, 1943, Strotz and Wold, 1960, Spirtes, Glymour, and Scheines, 1993, Pearl, 1993a, 2000a, Lindley, 2002) and has led to new ways of defining and estimating causal effects in nonlinear and nonparametric models (that is, models in which the functional form of the equations is unknown).

The central idea is to exploit the invariant characteristics of structural equations without committing to a specific functional form. For example, the nonparametric interpretation of the diagram of Fig. 2(a) corresponds to a set of three functions, each corresponding to one of the observed variables:

$$z = f_Z(u_Z)$$
$$x = f_X(z, u_X)$$
$$y = f_Y(x, u_Y) \quad (5)$$

where in this particular example U_Z, U_X and U_Y are assumed to be jointly independent but, otherwise, arbitrarily distributed. Each of these functions represents a causal process (or mechanism) that determines the value of the left variable (output) from those on the right variables (inputs). The absence of a variable from the right hand side of an equation encodes the assumption that Nature ignores that variable in the process of determining the value of the output variable. For example, the absence of variable Z from the arguments of f_Y conveys the empirical claim that variations in Z will leave Y unchanged, as long as variables U_Y, and X remain constant. A system of such functions are said to be *structural* if they are assumed to be autonomous, that is, each function is invariant to possible changes in the form of the other functions (Simon, 1953, Koopmans, 1953).

3.2.1. Representing interventions

This feature of invariance permits us to use structural equations as a basis for modeling causal effects and counterfactuals. This is done through a mathematical operator called *do(x)* which simulates physical interventions by deleting certain functions from the model, replacing them by a constant $X = x$, while keeping the rest of the model unchanged. For example, to emulate an intervention $do(x_0)$ that holds X constant (at $X = x_0$) in model M of Fig. 2(a), we replace the equation for x in Eq. (5) with $x = x_0$, and obtain a new model, M_{x0},

$$z = f_Z(u_Z)$$
$$x = x_0$$
$$y = f_Y(x, u_Y) \quad (6)$$

the graphical description of which is shown in Fig. 2(b).

The joint distribution associated with the modified model, denoted $P(z, y|do(x_0))$ describes the post-intervention distribution of variables Y and Z (also called "controlled" or "experimental" distribution), to be distinguished from the pre-intervention distribution, $P(x, y, z)$, associated with the original model of Eq. (5). For example, if X represents a treatment variable, Y a response variable, and Z some covariate that affects the amount of treatment received, then the distribution $P(z, y|do(x_0))$ gives the proportion of individuals that would attain response level $Y = y$ and covariate level $Z = z$ under the hypothetical situation in which treatment $X = x_0$ is administered uniformly to the population.

In general, we can formally define the post-intervention distribution by the equation:

(7) $P_M(y|do(x)) \triangleq P_{M_x}(y)$

In words: In the framework of model M, the post-intervention distribution of outcome Y is defined as the probability that model M_x assigns to each outcome level $Y = y$.

From this distribution, one is able to assess treatment efficacy by comparing aspects of this distribution at different levels of x_0. A common measure of treatment efficacy is the average difference

(8) $E(Y|do(x'_0)) - E(Y|do(x_0))$

where x'_0 and x_0 are two levels (or types) of treatment selected for comparison. Another measure is the experimental Risk Ratio

(9) $E(Y|do(x'_0))/E(Y|do(x_0))$.

The variance $Var(Y|do(x_0))$, or any other distributional parameter, may also enter the comparison; all these measures can be obtained from the controlled distribution function $P(Y = y|do(x)) = \sum_z P(z, y|do(x))$ which was called "causal effect" in Pearl (2000a, 1995) (see footnote 2). The central question in the analysis of causal effects is the question of *identification*: Can the controlled (post-intervention) distribution, $P(Y = y|do(x))$, be estimated from data governed by the pre-intervention distribution, $P(z, x, y)$?

The problem of *identification* has received considerable attention in econometrics (Hurwicz, 1950, Marschak, 1950, Koopmans, 1953) and social science (Duncan, 1975, Bollen, 1989), usually in linear parametric settings, where it reduces to asking whether some model parameter, β, has a unique solution in terms of the parameters of *P* (the distribution of the observed variables). In the nonparametric formulation, identification is more involved, since the notion of "has a unique solution" does not directly apply to causal quantities such as $Q(M) = P(y|do(x))$ which have no distinct parametric signature, and are defined procedurally by simulating an intervention in a causal model *M* (as in (6)). The following definition overcomes these difficulties:

Definition 2 (Identifiability (Pearl, 2000a, p. 77)) *A quantity Q(M) is identifiable, given a set of assumptions A, if for any two models M_1 and M_2 that satisfy A, we have*

(10) $P(M_1) = P(M_2) \Rightarrow Q(M_1) = Q(M_2)$

In words, the details of M_1 and M_2 do not matter; what matters is that the assumptions in *A* (e.g., those encoded in the diagram) would constrain the variability of those details in such a way that equality of *P*'s would entail equality of *Q*'s. When this happens, *Q* depends on *P* only, and should therefore be expressible in terms of the parameters of *P*. The next subsections exemplify and operationalize this notion.

3.2.2. Estimating the effect of interventions

To understand how hypothetical quantities such as $P(y|do(x))$ or $E(Y|do(x_0))$ can be estimated from actual data and a partially specified model let us begin with a simple demonstration on the model of Fig. 2(a). We will see that, despite our ignorance of f_X, f_Y, f_Z and $P(u)$, $E(Y|do(x_0))$ is nevertheless identifiable and is given by the conditional expectation $E(Y|X = x_0)$. We do this by deriving and comparing the expressions for these two quantities, as defined by (5) and (6), respectively. The mutilated model in Eq. (6) dictates:

(11) $E(Y|do(x_0)) = E(f_Y(x_0, u_Y))$,

whereas the pre-intervention model of Eq. (5) gives

$$E(Y|X=x_0)) = E(f_Y(X, u_Y)|X=x_0)$$
$$= E(f_Y(x_0, u_Y)|X=x_0)$$
$$= E(f_Y(x_0, u_Y))$$

(12)

which is identical to (11). Therefore,

(13) $E(Y|do(x_0)) = E(Y|X = x_0))$

Using a similar derivation, though somewhat more involved, we can show that $P(y|do(x))$ is identifiable and given by the conditional probability $P(y|x)$.

We see that the derivation of (13) was enabled by two assumptions; first, Y is a function of X and U_Y only, and, second, U_Y is independent of $\{U_Z, U_X\}$, hence of X. The latter assumption parallels the celebrated "orthogonality" condition in linear models, $Cov(X, U_Y) = 0$, which has been used routinely, often thoughtlessly, to justify the estimation of structural coefficients by regression techniques.

Naturally, if we were to apply this derivation to the linear models of Fig. 1(a) or 1(b), we would get the expected dependence between Y and the intervention $do(x_0)$:

$$E(Y|do(x_0)) = E(f_Y(x_0, u_Y))$$
$$= E(\beta x_0 + u_Y)$$
$$= \beta x_0$$

(14)

This equality endows β with its causal meaning as "effect coefficient." It is extremely important to keep in mind that in structural (as opposed to regressional) models, β is not "interpreted" as an effect coefficient but is "proven" to be one by the derivation above. β will retain this causal interpretation regardless of how X is actually selected (through the function f_X, Fig. 2(a)) and regardless of whether U_X and U_Y are correlated (as in Fig. 1(b)) or uncorrelated (as in Fig. 1(a)). Correlations may only impede our ability to estimate β from nonexperimental data, but will not change its definition as given in (14). Accordingly, and contrary to endless confusions in the literature (see footnote 12) structural equations say absolutely nothing about the conditional expectation $E(Y|X = x)$. Such connection may exist under special circumstances, e.g., if $cov(X, U_Y) = 0$, as in Eq. (13), but is otherwise irrelevant to the definition or interpretation of β as effect coefficient, or to the empirical claims of Eq. (1).

The next subsection will circumvent these derivations altogether by reducing the identification problem to a graphical procedure. Indeed, since graphs encode all the information that non-parametric structural equations represent, they should permit us to solve the identification problem without resorting to algebraic analysis.

3.2.3. Causal effects from data and graphs

Causal analysis in graphical models begins with the realization that all causal effects are identifiable whenever the model is *Markovian*, that is, the graph is acyclic (i.e., containing no directed cycles) and all the error terms are jointly independent. Non-Markovian models, such as those involving correlated errors (resulting from unmeasured confounders), permit identification only under certain conditions, and these conditions too can be determined from the graph structure (Section 3.3). The key to these results rests with the following basic theorem.

Theorem 1 (The Causal Markov Condition) *Any distribution generated by a Markovian model M can be factorized as:*

$$P(v_1, v_2, \ldots, v_n) = \prod_i P(v_i | pa_i)$$

(15)

where $V_1, V_2, \ldots, V_n$ *are the endogenous variables in M, and pa_i are (values of) the endogenous "parents" of V_i in the causal diagram associated with M.*

For example, the distribution associated with the model in Fig. 2(a) can be factorized as

(16) $P(z, y, x) = P(z)P(x|z)P(y|x)$

since *X* is the (endogenous) parent of *Y*, *Z* is the parent of *X*, and *Z* has no parents.

Corollary 1 (Truncated factorization) *For any Markovian model, the distribution generated by an intervention do(X = x₀) on a set X of endogenous variables is given by the truncated factorization*

(17)
$$P(v_1, v_2, \ldots, v_k | do(x_0)) = \prod_{i | V_i \notin X} P(v_i | pa_i)|_{x=x_0}$$

where *P(v_i|pa_i)* are the pre-intervention conditional probabilities.[5]

Corollary 1 instructs us to remove from the product of Eq. (15) those factors that quantify how the intervened variables (members of set *X*) are influenced by their pre-intervention parents. This removal follows from the fact that the post-intervention model is Markovian as well, hence, following Theorem 1, it must generate a distribution that is factorized according to the modified graph, yielding the truncated product of Corollary 1. In our example of Fig. 2(b), the distribution $P(z, y|do(x_0))$ associated with the modified model is given by

$P(z, y | do(x_0)) = P(z)P(y|x_0)$

where *P(z)* and *P(y|x₀)* are identical to those associated with the pre-intervention distribution of Eq. (16). As expected, the distribution of *Z* is not affected by the intervention, since

$$P(z|do(x_0)) = \sum_y P(z, y | do(x_0)) = \sum_y P(z)P(y|x_0) = P(z)$$

while that of *Y* is sensitive to *x₀*, and is given by

$$P(y|do)(x_0)) = \sum_z P(z, y|do(x_0)) = \sum_z P(z)P(y|x_0) = P(y|x_0)$$

This example demonstrates how the (causal) assumptions embedded in the model M permit us to predict the post-intervention distribution from the pre-intervention distribution, which further permits us to estimate the causal effect of X on Y from nonexperimental data, since $P(y|x_0)$ is estimable from such data. Note that we have made no assumption whatsoever on the form of the equations or the distribution of the error terms; it is the structure of the graph alone (specifically, the identity of X's parents) that permits the derivation to go through.

The truncated factorization formula enables us to derive causal quantities directly, without dealing with equations or equation modification as in Eqs. (11)–(13). Consider, for example, the model shown in Fig. 3, in which the error variables are kept implicit. Instead of writing down the corresponding five nonparametric equations, we can write the joint distribution directly as

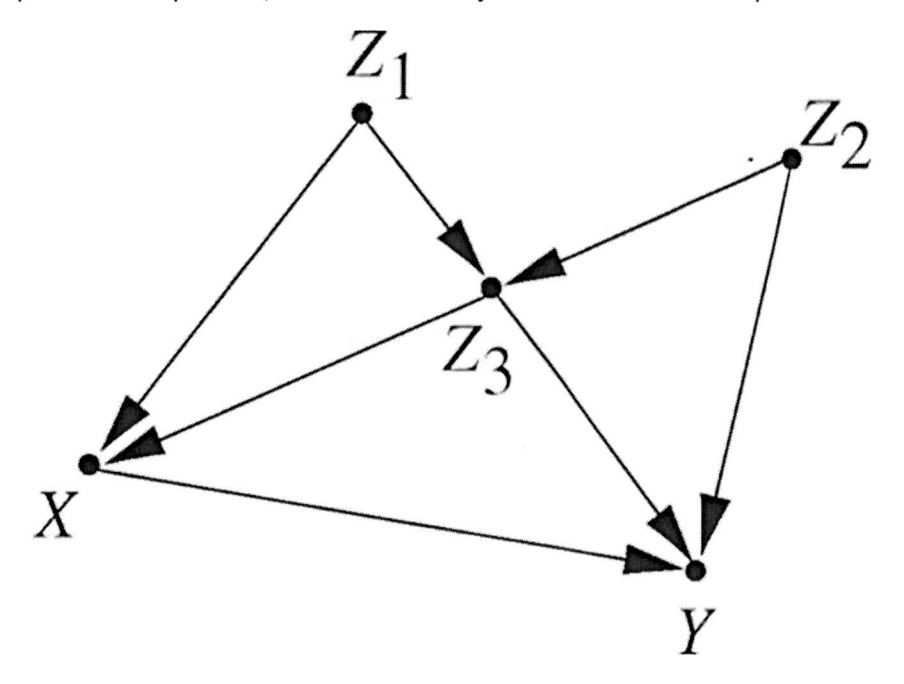

Figure 3: Markovian model illustrating the derivation of the causal effect of X on Y, Eq. (20). Error terms are not shown explicitly.

(18) $P(x, z_1, z_2, z_3, y) = P(z_1)P(z_2)P(z_3|z_1, z_2)P(x|z_1, z_3)P(y|z_2, z_3, x)$

where each marginal or conditional probability on the right hand side is directly estimable from the data. Now suppose we intervene and set variable X to x_0. The post-intervention distribution can readily be written (using the truncated factorization formula (17)) as

(19) $P(z_1, z_2, z_3, y|do(x_0)) = P(z_1)P(z_2)P(z_3|z_1, z_2)P(y|z_2, z_3, x_0)$

and the causal effect of X on Y can be obtained immediately by marginalizing over the Z variables, giving

(20)
$$P(y|do(x_0)) = \sum_{z_1, z_2, z_3} P(z_1)P(z_2)P(z_3|z_1, z_2)P(y|z_2, z_3, x_0)$$

Note that this formula corresponds precisely to what is commonly called "adjusting for Z_1, Z_2 and Z_3" and, moreover, we can write down this formula by inspection, without thinking on whether Z_1, Z_2 and Z_3 are confounders, whether they lie on the causal pathways, and so on. Though such questions can be answered explicitly from the topology of the graph, they are dealt with automatically when we write down the truncated factorization formula and marginalize.

Note also that the truncated factorization formula is not restricted to interventions on a single variable; it is applicable to simultaneous or sequential interventions such as those invoked in the analysis of time varying treatment with time varying confounders (Robins, 1986, Arjas and Parner, 2004). For example, if X and Z_2 are both treatment variables, and Z_1 and Z_3 are measured covariates, then the post-intervention distribution would be

(21) $P(z_1, z_3, y|do(x), do(z_2)) = P(z_1)P(z_3|z_1, z_2)P(y|z_2, z_3, x)$

and the causal effect of the treatment sequence $do(X = x)$, $do(Z_2 = z_2)$6 would be

(22)
$$P(y|do(x), do(z_2)) = \sum_{z_1, z_3} P(z_1)P(z_3|z_1, z_2)P(y|z_2, z_3, x)$$

This expression coincides with Robins' (1987) *G*-computation formula, which was derived from a more complicated set of (counterfactual) assumptions. As noted by Robins, the formula dictates an adjustment for covariates (e.g., Z_3) that might be affected by previous treatments (e.g., Z_2).

3.3. Coping with unmeasured confounders

Things are more complicated when we face unmeasured confounders. For example, it is not immediately clear whether the formula in Eq. (20) can be estimated if any of Z_1, Z_2 and Z_3 is not measured. A few but challenging algebraic steps would reveal that one can perform the summation over Z_2 to obtain

(23)
$$P(y|do(x_0)) = \sum_{z_1, z_3} P(z_1) P(z_3|z_1) P(y|z_1, z_3, x_0)$$

which means that we need only adjust for Z_1 and Z_3 without ever measuring Z_2. In general, it can be shown (Pearl, 2000a, p. 73) that, whenever the graph is Markovian the post-interventional distribution $P(Y = y | do(X = x))$ is given by the following expression:

$$P(Y=y|do(X=x)) = \sum_{t} P(y|t, x) P(t)$$
(24)

where *T* is the set of direct causes of *X* (also called "parents") in the graph. This allows us to write (23) directly from the graph, thus skipping the algebra that led to (23). It further implies that, no matter how complicated the model, the parents of *X* are the only variables that need to be measured to estimate the causal effects of *X*.

It is not immediately clear however whether other sets of variables beside *X*'s parents suffice for estimating the effect of *X*, whether some algebraic manipulation can further reduce Eq. (23), or that measurement of Z_3 (unlike Z_1, or Z_2) is necessary in any estimation of $P(y|do(x_0))$. Such considerations become transparent from a graphical criterion to be discussed next.

3.3.1. Covariate selection – the back-door criterion

Consider an observational study where we wish to find the effect of X on Y, for example, treatment on response, and assume that the factors deemed relevant to the problem are structured as in Fig. 4; some are affecting the response, some are affecting the treatment and some are affecting both treatment and response. Some of these factors may be unmeasurable, such as genetic trait or life style, others are measurable, such as gender, age, and salary level. Our problem is to select a subset of these factors for measurement and adjustment, namely, that if we compare treated vs. untreated subjects having the same values of the selected factors, we get the correct treatment effect in that subpopulation of subjects. Such a set of factors is called a "sufficient set" or "admissible set" for adjustment. The problem of defining an admissible set, let alone finding one, has baffled epidemiologists and social scientists for decades (see (Greenland et al., 1999, Pearl, 1998) for review).

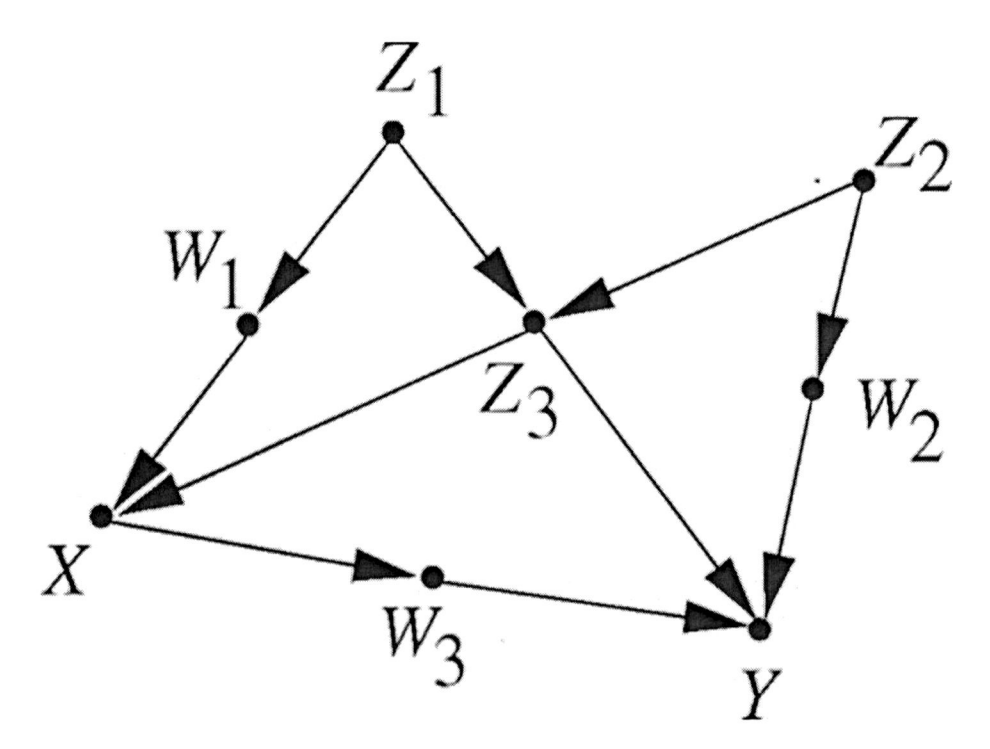

Figure 4: Markovian model illustrating the back-door criterion. Error terms are not shown explicitly.

The following criterion, named "back-door" in (Pearl, 1993a), settles this problem by providing a graphical method of selecting admissible sets of factors for adjustment.

Definition 3 (Admissible sets – the back-door criterion) *A set S is admissible (or "sufficient") for adjustment if two conditions hold:*

No element of S is a descendant of X

The elements of S "block" all "back-door" paths from X to Y, namely all paths that end with an arrow pointing to X.

In this criterion, "blocking" is interpreted as in Definition 1. For example, the set $S = \{Z_3\}$ blocks the path $X \leftarrow W_1 \leftarrow Z_1 \rightarrow Z_3 \rightarrow Y$, because the arrow-emitting node Z_3 is in S. However, the set $S = \{Z_3\}$ does not block the path $X \leftarrow W_1 \leftarrow Z_1 \rightarrow Z_3 \leftarrow Z_2 \rightarrow W_2 \rightarrow Y$, because none of the arrow-emitting nodes, Z_1 and Z_2, is in S, and the collision node Z_3 is not outside S.

Based on this criterion we see, for example, that the sets $\{Z_1, Z_2, Z_3\}$, $\{Z_1, Z_3\}$, $\{W_1, Z_3\}$, and $\{W_2, Z_3\}$, each is sufficient for adjustment, because each blocks all back-door paths between X and Y. The set $\{Z_3\}$, however, is not sufficient for adjustment because, as explained above, it does not block the path $X \leftarrow W_1 \leftarrow Z_1 \rightarrow Z_3 \leftarrow Z_2 \rightarrow W_2 \rightarrow Y$.

The intuition behind the back-door criterion is as follows. The back-door paths in the diagram carry spurious associations from X to Y, while the paths directed along the arrows from X to Y carry causative associations. Blocking the former paths (by conditioning on S) ensures that the measured association between X and Y is purely causative, namely, it correctly represents the target quantity: the causal effect of X on Y. The reason for excluding descendants of X (e.g., W_3 or any of its descendants) is given in (Pearl, 2009b, pp. 338–41).

Formally, the implication of finding an admissible set S is that, stratifying on S is guaranteed to remove all confounding bias relative the causal effect of X on Y. In other words, the risk difference in each stratum of S gives the correct causal effect in that stratum. In the binary case, for example, the risk difference in stratum s of S is given by

$P(Y = 1 | X = 1, S = s) - P(Y = 1 | X = 0, S = s)$

while the causal effect (of X on Y) at that stratum is given by

$P(Y = 1 | do(X = 1), S = s) - P(Y = 1 | do(X = 0), S = s)$.

These two expressions are guaranteed to be equal whenever S is a sufficient set, such as $\{Z_1, Z_3\}$ or $\{Z_2, Z_3\}$ in Fig. 4. Likewise, the average stratified risk difference, taken over all strata,

$$\sum_s [P(Y=1|X=1, S=s) - P(Y=1|X=0, S=s)]P(S=s),$$

gives the correct causal effect of X on Y in the entire population

$P(Y = 1|do(X = 1)) - P(Y = 1|do(X = 0))$.

In general, for multi-valued variables X and Y, finding a sufficient set S permits us to write

$P(Y = y|do(X = x), S = s) = P(Y = y|X = x, S = s)$

and

(25)
$$P(Y=y|do(X=x)) = \sum_s P(Y=y|X=x, S=s)P(S=s)$$

Since all factors on the right hand side of the equation are estimable (e.g., by regression) from the pre-interventional data, the causal effect can likewise be estimated from such data without bias.

An equivalent expression for the causal effect (25) can be obtained by multiplying and dividing by the conditional probability $P(X = x|S = s)$, giving

(26)
$$P(Y=y|do(X=x)) = \sum_s \frac{P(Y=y, X=x, S=s)}{P(X=x|S=s)}$$

from which the name "Inverse Probability Weighting" has evolved (Pearl, 2000a, pp. 73, 95).

Interestingly, it can be shown that any irreducible sufficient set, S, taken as a unit, satisfies the associational criterion that epidemiologists have been using to define "confounders". In other words, S must be associated with X and, simultaneously, associated with Y, given X. This need not hold for any specific members of S. For example, the variable Z_3 in Fig. 4, though it is a member of every sufficient set and hence a confounder, can be unassociated with both Y and X (Pearl, 2000a, p. 195). Conversely, a pre-treatment variable Z that is associated with both Y and X may need to be excluded from entering a sufficient set.

The back-door criterion allows us to write Eq. (25) directly, by selecting a sufficient set S directly from the diagram, without manipulating the truncated factorization formula. The selection criterion can be applied systematically to diagrams of any size and shape, thus freeing analysts from judging whether "X is conditionally ignorable given S," a formidable mental task required in the potential-response framework (Rosenbaum and Rubin, 1983). The criterion also enables the analyst to search for an optimal set of covariate—namely, a set S that minimizes measurement cost or sampling variability (Tian, Paz, and Pearl, 1998).

All in all, one can safely state that, armed with the back-door criterion, causality has removed "confounding" from its store of enigmatic and controversial concepts.

3.3.2. Confounding equivalence – a graphical test

Another problem that has been given graphical solution recently is that of determining whether adjustment for two sets of covariates would result in the same confounding bias (Pearl and Paz, 2009). The reasons for posing this question are several. First, an investigator may wish to assess, prior to taking any measurement, whether two candidate sets of covariates, differing substantially in dimensionality, measurement error, cost, or sample variability are equally valuable in their bias-reduction potential. Second, assuming that the structure of the underlying DAG is only partially known, one may wish to test, using adjustment, which of two hypothesized structures is compatible with the data. Structures that predict equal response to adjustment for two sets of variables must be rejected if, after adjustment, such equality is not found in the data.

Definition 4 ((c-equivalence)) *Define two sets, T and Z of covariates as c-equivalent, (c connotes "confounding") if the following equality holds:*

$$\sum_t P(y|x,t)P(t) = \sum_z P(y|x,z)P(z) \qquad \forall x,y \qquad (27)$$

Definition 5 ((Markov boundary)) *For any set of variables S in a DAG G, the Markov boundary S_m of S is the minimal subset of S that d-separates X from all other members of S.*

In Fig. 4, for example, the Markov boundary of $S = \{W_1, Z_1, Z_2, Z_3\}$ is $S_m = \{W_1, Z_3\}$.

Theorem 2 *(Pearl and Paz, 2009)*

Let Z and T be two sets of variables in G, containing no descendant of X. A necessary and sufficient conditions for Z and T to be c-equivalent is that at least one of the following conditions holds:

$Z_m = T_m$, (i.e., the Markov boundary of Z coincides with that of T)

Z and T are admissible (i.e., satisfy the back-door condition)

For example, the sets $T = \{W_1, Z_3\}$ and $Z = \{Z_3, W_2\}$ in Fig. 4 are c-equivalent, because each blocks all back-door paths from X to Y. Similarly, the non-admissible sets $T = \{Z_2\}$ and $Z = \{W_2, Z_2\}$ are c-equivalent, since their Markov boundaries are the same ($T_m = Z_m = \{Z_2\}$). In contrast, the sets $\{W_1\}$ and $\{Z_1\}$, although they block the same set of paths in the graph, are not c-equivalent; they fail both conditions of Theorem 2.

Tests for c-equivalence (27) are fairly easy to perform, and they can also be assisted by propensity scores methods. The information that such tests provide can be as powerful as conditional independence tests. The statistical ramification of such tests are explicated in (Pearl and Paz, 2009).

3.3.3. General control of confounding

Adjusting for covariates is only one of many methods that permits us to estimate causal effects in nonexperimental studies. Pearl (1995) has presented examples in which there exists no set of variables that is sufficient for adjustment and where the causal effect can nevertheless be estimated consistently. The estimation, in such cases, employs multi-stage adjustments.

For example, if W_3 is the only observed covariate in the model of Fig. 4, then there exists no sufficient set for adjustment (because no set of observed covariates can block the paths from X to Y through Z_3), yet $P(y|do(x))$ can be estimated in two steps; first we estimate $P(w_3|do(x)) = P(w_3|x)$ (by virtue of the fact that there exists no unblocked back-door path from X to W_3), second we estimate $P(y|do(w_3))$ (since X constitutes a sufficient set for the effect of W_3 on Y) and, finally, we combine the two effects together and obtain

(28)
$$P(y|do(x)) = \sum_{w_3} P(w_3|do(x))P(y|do(w_3))$$

In this example, the variable W_3 acts as a "mediating instrumental variable" (Pearl, 1993b, Chalak and White, 2006).

The analysis used in the derivation and validation of such results invokes mathematical rules of transforming causal quantities, represented by expressions such as $P(Y = y|do(x))$, into *do*-free expressions derivable from $P(z, x, y)$, since only *do*-free expressions are estimable from non-experimental data. When such a transformation is feasible, we are ensured that the causal quantity is identifiable.

Applications of this calculus to problems involving multiple interventions (e.g., time varying treatments), conditional policies, and surrogate experiments were developed in Pearl and Robins (1995), Kuroki and Miyakawa (1999), and Pearl (2000a, Chapters 3–4).

A more recent analysis (Tian and Pearl, 2002) shows that the key to identifiability lies not in blocking paths between X and Y but, rather, in blocking paths between X and its immediate successors on the pathways to Y. All existing criteria for identification are special cases of the one defined in the following theorem:

Theorem 3 (Tian and Pearl, 2002) *A sufficient condition for identifying the causal effect $P(y|do(x))$ is that every path between X and any of its children traces at least one arrow emanating from a measured variable.*[7]

For example, if W_3 is the only observed covariate in the model of Fig. 4,

$P(y|do(x))$ can be estimated since every path from X to W_3 (the only child of X) traces either the arrow $X \to W_3$, or the arrow $W_3 \to Y$, both emanating from a measured variable (W_3).

Shpitser and Pearl (2006) have further extended this theorem by (1) presenting a *necessary* and sufficient condition for identification, and (2) extending the condition from causal effects to any counterfactual expression. The corresponding unbiased estimands for these causal quantities are readable directly from the diagram.

Graph-based methods for effect identification under measurement errors are discussed in (Pearl, 2009f, Hernán and Cole, 2009, Cai and Kuroki, 2008).

3.3.4. From identification to estimation

The mathematical derivation of causal effect estimands, like Eqs. (25) and (28) is merely a first step toward computing quantitative estimates of those effects from finite samples, using the rich traditions of statistical estimation and machine learning Bayesian as well as non-Bayesian. Although the estimands derived in (25) and (28) are non-parametric, this does not mean that one should refrain from using parametric forms in the estimation phase of the study. Parameterization is in fact necessary when the dimensionality of a problem is high. For example, if the assumptions of Gaussian, zero-mean disturbances and additive interactions are deemed reasonable, then the estimand given in (28) can be converted to the product $E(Y|do(x)) = r_{W_3X} r_{YW_3 \cdot X} \, x$, where $r_{YZ \cdot X}$ is the (standardized) coefficient of Z in the regression of Y on Z and X. More sophisticated estimation techniques are the "marginal structural models" of (Robins, 1999), and the "propensity score" method of (Rosenbaum and Rubin, 1983) which were found to be particularly useful when dimensionality is high and data are sparse (see Pearl (2009b, pp. 348–52)).

It should be emphasized, however, that contrary to conventional wisdom (e.g., (Rubin, 2007, 2009)), propensity score methods are merely efficient estimators of the right hand side of (25); they entail the same asymptotic bias, and cannot be expected to reduce bias in case the set S does not satisfy the back-door criterion (Pearl, 2000a, 2009c,d). Consequently, the prevailing practice of conditioning on as many pre-treatment measurements as possible should be approached with great caution; some covariates (e.g., Z_3 in Fig. 3) may actually increase bias if included in the analysis (see footnote 16). Using simulation and parametric analysis, Heckman and Navarro-Lozano (2004) and Wooldridge (2009) indeed confirmed the bias-raising potential of certain covariates in propensity-score methods. The graphical tools presented in this section unveil the character of these covariates and show precisely what covariates should, and should not be included in the conditioning set for propensity-score matching (see also (Pearl and Paz, 2009, Pearl, 2009e)).

3.4. Counterfactual analysis in structural models

Not all questions of causal character can be encoded in $P(y|do(x))$ type expressions, thus implying that not all causal questions can be answered from experimental studies. For example, questions of attribution (e.g., what fraction of death cases are *due to* specific exposure?) or of susceptibility (what fraction of the healthy unexposed population would have gotten the disease had they been exposed?) cannot be answered from experimental studies, and naturally, this kind of questions cannot be expressed in $P(y|do(x))$ notation.8 To answer such questions, a probabilistic analysis of counterfactuals is required, one dedicated to the relation "Y would be y had X been x in situation $U = u$," denoted $Y_x(u) = y$. Remarkably, unknown to most economists and philosophers, structural equation models provide the formal interpretation and symbolic machinery for analyzing such counterfactual relationships.9

The key idea is to interpret the phrase "had X been x" as an instruction to make a minimal modification in the current model, which may have assigned X a different value, say $X = x'$, so as to ensure the specified condition $X = x$. Such a minimal modification amounts to replacing the equation for X by a constant x, as we have done in Eq. (6). This replacement permits the constant x to differ from the actual value of X (namely $f_X(z, u_X)$) without rendering the system of equations inconsistent, thus yielding a formal interpretation of counterfactuals in multi-stage models, where the dependent variable in one equation may be an independent variable in another.

Definition 6 (Unit-level Counterfactuals – "surgical" definition, Pearl (2000a, p. 98)) *Let M be a structural model and M_x a modified version of M, with the equation(s) of X replaced by $X = x$. Denote the solution for Y in the equations of M_x by the symbol $Y_{Mx}(u)$. The counterfactual $Y_x(u)$ (Read: "The value of Y in unit u, had X been x") is given by:*

(29) $Y_x(u) \triangleq Y_{Mx}(u).$

In words: The counterfactual $Y_x(u)$ in model M is defined as the solution for Y in the "surgically modified" submodel M_x.

We see that the unit-level counterfactual $Y_x(u)$, which in the Neyman-

Rubin approach is treated as a primitive, undefined quantity, is actually a derived quantity in the structural framework. The fact that we equate the experimental unit u with a vector of background conditions, $U = u$, in M, reflects the understanding that the name of a unit or its identity do not matter; it is only the vector $U = u$ of attributes characterizing a unit which determines its behavior or response. As we go from one unit to another, the laws of nature, as they are reflected in the functions f_X, f_Y, etc. remain invariant; only the attributes $U = u$ vary from individual to individual.[10]

To illustrate, consider the solution of Y in the modified model M_{x0} of Eq. (6), which Definition 6 endows with the symbol $Y_{x0}(u_X, u_Y, u_Z)$. This entity has a clear counterfactual interpretation, for it stands for the way an individual with characteristics (u_X, u_Y, u_Z) would respond, had the treatment been x_0, rather than the treatment $x = f_X(z, u_X)$ actually received by that individual. In our example, since Y does not depend on u_X and u_Z, we can write:

(30) $Y_{x0}(u) = Y_{x0}(u_Y, u_X, u_Z) = f_Y(x_0, u_Y)$.

In a similar fashion, we can derive

$Y_{z0}(u) = f_Y(f_X(z_0, u_X), u_Y)$,

$X_{z0,y0}(u) = f_X(z_0, u_X)$,

and so on. These examples reveal the counterfactual reading of each individual structural equation in the model of Eq. (5). The equation $x = f_X(z, u_X)$, for example, advertises the empirical claim that, regardless of the values taken by other variables in the system, had Z been z_0, X would take on no other value but $x = f_X(z_0, u_X)$.

Clearly, the distribution $P(u_Y, u_X, u_Z)$ induces a well defined probability on the counterfactual event $Y_{x0} = y$, as well as on joint counterfactual events, such as '$Y_{x0} = y$ AND $Y_{x1} = y'$,' which are, in principle, unobservable if $x_0 \neq x_1$. Thus, to answer attributional questions, such as whether Y would be y_1 if X were x_1, given that in fact Y is y_0 and X is x_0, we need to compute the conditional probability $P(Y_{x1} = y_1 | Y = y_0, X = x_0)$ which is well defined once we know the forms of the structural equations and the distribution of the exogenous variables in the model. For example, assuming linear equations (as in Fig. 1),

$x = u_X \qquad y = \beta x + u_X$,

the conditioning events $Y = y_0$ and $X = x_0$ yield $U_X = x_0$ and $U_Y = y_0 - \beta x_0$, and we can conclude that, with probability one, Y_{x_1} must take on the value: $Y_{x_1} = \beta x_1 + U_Y = \beta(x_1 - x_0) + y_0$. In other words, if X were x_1 instead of x_0, Y would increase by β times the difference $(x_1 - x_0)$. In nonlinear systems, the result would also depend on the distribution of $\{U_X, U_Y\}$ and, for that reason, attributional queries are generally not identifiable in nonparametric models (see Section 6.3 and 2000a, Chapter 9).

In general, if x and x' are incompatible then Y_x and $Y_{x'}$ cannot be measured simultaneously, and it may seem meaningless to attribute probability to the joint statement "Y would be y if $X = x$ and Y would be y' if $X = x'$."[11] Such concerns have been a source of objections to treating counterfactuals as jointly distributed random variables (Dawid, 2000). The definition of Y_x and $Y_{x'}$ in terms of two distinct submodels neutralizes these objections (Pearl, 2000b), since the contradictory joint statement is mapped into an ordinary event, one where the background variables satisfy both statements simultaneously, each in its own distinct submodel; such events have well defined probabilities.

The surgical definition of counterfactuals given by (29), provides the conceptual and formal basis for the Neyman-Rubin potential-outcome framework, an approach to causation that takes a controlled randomized trial (CRT) as its ruling paradigm, assuming that nothing is known to the experimenter about the science behind the data. This "black-box" approach, which has thus far been denied the benefits of graphical or structural analyses, was developed by statisticians who found it difficult to cross the two mental barriers discussed in Section 2.2. Section 5 establishes the precise relationship between the structural and potential-outcome paradigms, and outlines how the latter can benefit from the richer representational power of the former.

4. Methodological Principles of Causal Inference

The structural theory described in the previous sections dictates a principled methodology that eliminates much of the confusion concerning the interpretations of study results as well as the ethical dilemmas that this confusion tends to spawn. The methodology dictates that every investigation involving causal relationships (and this entails the vast majority of empirical studies in the health, social, and behavioral sciences) should be structured along the following four-step process:

Define: Express the target quantity Q as a function $Q(M)$ that can be computed from any model M.

Assume: Formulate causal assumptions using ordinary scientific language and represent their structural part in graphical form.

Identify: Determine if the target quantity is identifiable (i.e., expressible in terms of estimable parameters).

Estimate: Estimate the target quantity if it is identifiable, or approximate it, if it is not. Test the statistical implications of the model, if any, and modify the model when failure occurs.

4.1. Defining the target quantity

The definitional phase is the most neglected step in current practice of quantitative analysis. The structural modeling approach insists on defining the target quantity, be it "causal effect," "mediated effect," "effect on the treated," or "probability of causation" before specifying any aspect of the model, without making functional or distributional assumptions and prior to choosing a method of estimation.

The investigator should view this definition as an *algorithm* that receives a model M as an input and delivers the desired quantity $Q(M)$ as the output. Surely, such algorithm should not be tailored to any aspect of the input M; it should be general, and ready to accommodate any conceivable model M whatsoever. Moreover, the investigator should imagine that the input M is a completely specified model, with all the functions f_X, f_Y, . . . and all the U variables (or their associated probabilities) given precisely. This is the hardest step for statistically trained investigators to make; knowing in advance that such model details will never be estimable from the data, the definition of $Q(M)$ appears like a futile exercise in fantasy land – it is not.

For example, the formal definition of the causal effect $P(y|do(x))$, as given in Eq. (7), is universally applicable to all models, parametric as well as nonparametric, through the formation of a submodel M_x. By defining causal effect procedurally, thus divorcing it from its traditional parametric representation, the structural theory avoids the many pitfalls and confusions that have plagued the interpretation of structural and regressional parameters for the past half century.12

4.2. Explicating causal assumptions

This is the second most neglected step in causal analysis. In the past, the difficulty has been the lack of a language suitable for articulating causal assumptions which, aside from impeding investigators from explicating assumptions, also inhibited them from giving causal interpretations to their findings.

Structural equation models, in their counterfactual reading, have removed this lingering difficulty by providing the needed language for causal analysis. Figures 3 and 4 illustrate the graphical component of this language, where assumptions are conveyed through the missing arrows in the diagram. If numerical or functional knowledge is available, for example, linearity or monotonicity of the functions $f_X, f_Y, \ldots$, those are stated separately, and applied in the identification and estimation phases of the study. Today we understand that the longevity and natural appeal of structural equations stem from the fact that they permit investigators to communicate causal assumptions formally and in the very same vocabulary in which scientific knowledge is stored.

Unfortunately, however, this understanding is not shared by all causal analysts; some analysts vehemently oppose the re-emergence of structure-based causation and insist, instead, on articulating causal assumptions exclusively in the unnatural (though formally equivalent) language of "potential outcomes," "ignorability," "missing data," "treatment assignment," and other metaphors borrowed from clinical trials. This modern assault on structural models is perhaps more dangerous than the regressional invasion that distorted the causal readings of these models in the late 1970s (Richard, 1980). While sanctioning causal inference in one idiosyncratic style of analysis, the modern assault denies validity to any other style, including structural equations, thus discouraging investigators from subjecting models to the scrutiny of scientific knowledge.

This exclusivist attitude is manifested in passages such as: "The crucial idea is to set up the causal inference problem as one of missing data" or "If a problem of causal inference cannot be formulated in this manner (as the comparison of potential outcomes under different treatment assignments), it is not a problem of inference for causal effects, and the use of "causal" should be avoided," or, even more bluntly, "the underlying assumptions needed to justify any causal conclusions should be carefully and explicitly argued, not in terms of technical properties like "uncorrelated error terms," but in terms of real world properties, such as how the units received the different treatments" (Wilkinson, the Task Force on Statistical Inference, and *APA Board of Scientific Affairs*, 1999).

The methodology expounded in this paper testifies against such restrictions. It demonstrates the viability and scientific soundness of the traditional structural equations paradigm, which stands diametrically opposed to the "missing data" paradigm. It renders the vocabulary of "treatment assignment" stifling and irrelevant (e.g., there is no "treatment assignment" in sex discrimination cases). Most importantly, it strongly prefers the use of "uncorrelated error terms," (or "omitted factors") over its "strong ignorability" alternative, as the proper way of articulating causal assumptions. Even the most devout advocates of the "strong ignorability" language use "omitted factors" when the need arises to defend assumptions (e.g., (Sobel, 2008))

4.3. Identification, estimation, and approximation

Having unburden itself from parametric representations, the identification process in the structural framework proceeds either in the space of assumptions (i.e., the diagram) or in the space of mathematical expressions, after translating the graphical assumptions into a counterfactual language, as demonstrated in Section 5.3. Graphical criteria such as those of Definition 3 and Theorem 3 permit the identification of causal effects to be decided entirely within the graphical domain, where it can benefit from the guidance of scientific understanding. Identification of counterfactual queries, on the other hand, often require a symbiosis of both algebraic and graphical techniques. The nonparametric nature of the identification task (Definition 1) makes it clear that contrary to traditional folklore in linear analysis, it is not the model that need be identified but the query Q – the target of investigation. It also provides a simple way of proving non-identifiability: the construction of two parameterization of M, agreeing in P and disagreeing in Q, is sufficient to rule out identifiability.

When Q is identifiable, the structural framework also delivers an algebraic expression for the estimand $EST(Q)$ of the target quantity Q, examples of which are given in Eqs. (24) and (25), and estimation techniques are then unleashed as discussed in Section 3.3.4. An integral part of this estimation phase is a test for the testable implications, if any, of those assumptions in M that render Q identifiable – there is no point in estimating $EST(Q)$ if the data proves those assumptions false and $EST(Q)$ turns out to be a misrepresentation of Q. Investigators should be reminded, however, that only a fraction, called "kernel," of the assumptions embodied in M are needed for identifying Q (Pearl, 2004), the rest may be violated in the data with no effect on Q. In Fig. 2, for example, the assumption $\{U_Z \perp\!\!\!\perp U_X\}$ is not necessary for identifying $Q = P(y|do(x))$; the kernel $\{U_Y \perp\!\!\!\perp U_Z, U_Y \perp\!\!\!\perp U_X\}$ (together with the missing arrows) is sufficient. Therefore, the testable implication of this kernel, $Z \perp\!\!\!\perp Y | X$, is all we need to test when our target quantity is Q; the assumption $\{U_Z \perp\!\!\!\perp U_X\}$ need not concern us.

More importantly, investigators must keep in mind that only a tiny fraction of any kernel lends itself to statistical tests, the bulk of it must remain untestable, at the mercy of scientific judgment. In Fig. 2, for example, the assumption set $\{U_X \perp\!\!\!\perp U_Z, U_Y \perp\!\!\!\perp U_X\}$ constitutes a sufficient kernel for $Q = P(y|do(x))$ (see Eq. (28)) yet it has no testable implications whatsoever. The prevailing practice of submitting an entire structural equation model to a "goodness of fit" test (Bollen, 1989) in support of causal claims is at odd with the logic of SCM (see (Pearl, 2000a, pp. 144–5)). Alternative causal models usually exist that make contradictory claims and, yet, possess identical statistical implications. Statistical test can be used for rejecting certain kernels, in the rare cases where such kernels have testable implications, but the lion's share of supporting causal claims falls on the shoulders of untested causal assumptions.

When conditions for identification are not met, the best one can do is derive *bounds* for the quantities of interest—namely, a range of possible values of Q that represents our ignorance about the details of the data-generating process M and that cannot be improved with increasing sample size. A classical example of non identifiable model that has been approximated by bounds, is the problem of estimating causal effect in experimental studies marred by non compliance, the structure of which is given in Fig. 5.

Our task in this example is to find the highest and lowest values of Q

$$Q \triangleq P(Y=y|do(x)) = \sum_{u_X} P(Y=y|X=x, U_X=u_X)P(U_X=u_X)$$
(31)

subject to the equality constraints imposed by the observed probabilities $P(x, y, |z)$, where the maximization ranges over all possible functions $P(u_Y, u_X)$, $P(y|x, u_X)$ and $P(x|z, u_Y)$ that satisfy those constraints.

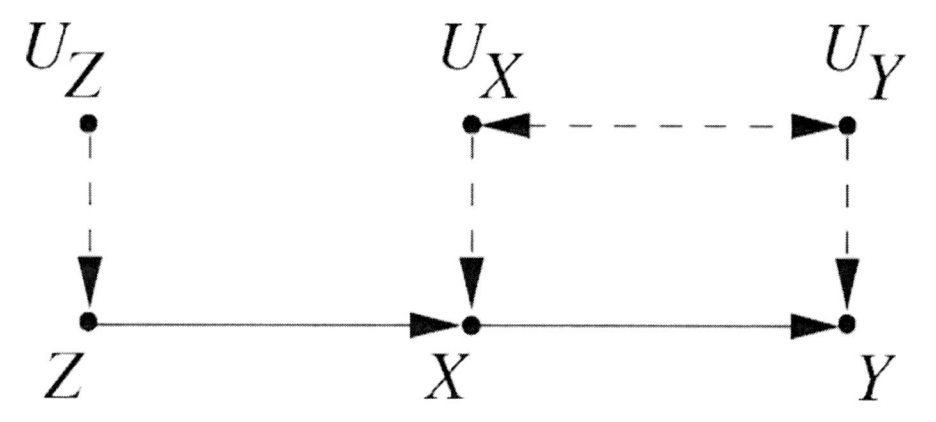

Figure 5: Causal diagram representing the assignment (Z), treatment (X), and outcome (Y) in a clinical trial with imperfect compliance.

Realizing that units in this example fall into 16 equivalence classes, each representing a binary function $X = f(z)$ paired with a binary function $y = g(x)$, Balke and Pearl (1997) were able to derive closed-form solutions for these bounds.13 They showed that, in certain cases, the derived bounds can yield significant information on the treatment efficacy. Chickering and Pearl (1997) further used Bayesian techniques (with Gibbs sampling) to investigate the sharpness of these bounds as a function of sample size. Kaufman, Kaufman, and MacLenose (2009) used this technique to bound direct and indirect effects (see Section 6.1).

5. The Potential Outcome Framework

This section compares the structural theory presented in Sections 1–3 to the potential-outcome framework, usually associated with the names of Neyman (1923) and Rubin (1974), which takes the randomized experiment as its ruling paradigm and has appealed therefore to researchers who do not find that paradigm overly constraining. This framework is not a contender for a comprehensive theory of causation for it is subsumed by the structural theory and excludes ordinary cause-effect relationships from its assumption vocabulary. We here explicate the logical foundation of the Neyman-Rubin framework, its formal subsumption by the structural causal model, and how it can benefit from the insights provided by the broader perspective of the structural theory.

The primitive object of analysis in the potential-outcome framework is the unit-based response variable, denoted $Y_x(u)$, read: "the value that outcome Y would obtain in experimental unit u, had treatment X been x." Here, *unit* may stand for an individual patient, an experimental subject, or an agricultural plot. In Section 3.4 (Eq. (29) we saw that this counterfactual entity has a natural interpretation in the SCM; it is the solution for Y in a modified system of equations, where *unit* is interpreted a vector u of background factors that characterize an experimental unit. Each structural equation model thus carries a collection of assumptions about the behavior of hypothetical units, and these assumptions permit us to derive the counterfactual quantities of interest. In the potential-outcome framework, however, no equations are available for guidance and $Y_x(u)$ is taken as primitive, that is, an undefined quantity in terms of which other quantities are defined; not a quantity that can be derived *from* the model. In this sense the structural interpretation of $Y_x(u)$ given in (29) provides the formal basis for the potential-outcome approach; the formation of the submodel M_x explicates mathematically how the hypothetical condition "had X been x" is realized, and what the logical consequences are of such a condition.

5.1. The "black-box" missing-data paradigm

The distinct characteristic of the potential-outcome approach is that, although investigators must think and communicate in terms of undefined, hypothetical quantities such as $Y_x(u)$, the analysis itself is conducted almost entirely within the axiomatic framework of probability theory. This is accomplished, by postulating a "super" probability function on both hypothetical and real events. If U is treated as a random variable then the value of the counterfactual $Y_x(u)$ becomes a random variable as well, denoted as Y_x.

The potential-outcome analysis proceeds by treating the observed distribution $P(x_1, \ldots, x_n)$ as the marginal distribution of an augmented probability function P^* defined over both observed and counterfactual variables. Queries about causal effects (written $P(y|do(x))$ in the structural analysis) are phrased as queries about the marginal distribution of the counterfactual variable of interest, written $P^*(Y_x = y)$. The new hypothetical entities Y_x are treated as ordinary random variables; for example, they are assumed to obey the axioms of probability calculus, the laws of conditioning, and the axioms of conditional independence.

Naturally, these hypothetical entities are not entirely whimsy. They are assumed to be connected to observed variables via consistency constraints (Robins, 1986) such as

(32) $X = x \Rightarrow Y_x = Y,$

which states that, for every u, if the actual value of X turns out to be x, then the value that Y would take on if 'X were x' is equal to the actual value of Y. For example, a person who chose treatment x and recovered, would also have recovered if given treatment x by design. When X is binary, it is sometimes more convenient to write (32) as:

$Y = xY_1 + (1 - x)Y_0$

Whether additional constraints should tie the observables to the unobservables is not a question that can be answered in the potential-outcome framework; for it lacks an underlying model to define its axioms.

The main conceptual difference between the two approaches is that, whereas the structural approach views the intervention $do(x)$ as an operation that changes a distribution but keeps the variables the same, the potential-outcome approach views the variable Y under $do(x)$ to be a different variable, Y_x, loosely connected to Y through relations such as (32), but remaining unobserved whenever $X \neq x$. The problem of inferring probabilistic properties of Y_x, then becomes one of "missing-data" for which estimation techniques have been developed in the statistical literature.

Pearl (2000a, Chapter 7) shows, using the structural interpretation of $Y_x(u)$, that it is indeed legitimate to treat counterfactuals as jointly distributed random variables in all respects, that consistency constraints like (32) are automatically satisfied in the structural interpretation and, moreover, that investigators need not be concerned about any additional constraints except the following two

(33) $Y_{yz} = y$ for all y, subsets Z, and values z for Z

(34) $X_z = x \Rightarrow Y_{xz} = Y_z$ for all x, subsets Z, and values z for Z

Equation (33) ensures that the interventions $do(Y = y)$ results in the condition $Y = y$, regardless of concurrent interventions, say $do(Z = z)$, that may be applied to variables other than Y. Equation (34) generalizes (32) to cases where Z is held fixed, at z. (See (Halpern, 1998) for proof of completeness.)

5.2. Problem formulation and the demystification of "ignorability"

The main drawback of this black-box approach surfaces in problem formulation, namely, the phase where a researcher begins to articulate the "science" or "causal assumptions" behind the problem of interest. Such knowledge, as we have seen in Section 1, must be articulated at the onset of every problem in causal analysis – causal conclusions are only as valid as the causal assumptions upon which they rest.

To communicate scientific knowledge, the potential-outcome analyst must express assumptions as constraints on P^*, usually in the form of conditional independence assertions involving counterfactual variables. For instance, in our example of Fig. 5, to communicate the understanding that Z is randomized (hence independent of U_X and U_Y), the potential-outcome analyst would use the independence constraint $Z \perp\!\!\!\perp \{Y_{z1}, Y_{z2}, \ldots, Y_{zk}\}$.[14] To further formulate the understanding that Z does not affect Y directly, except through X, the analyst would write a, so called, "exclusion restriction": $Y_{xz} = Y_x$.

A collection of constraints of this type might sometimes be sufficient to permit a unique solution to the query of interest. For example, if one can plausibly assume that, in Fig. 4, a set Z of covariates satisfies the conditional independence

(35) $Y_x \perp\!\!\!\perp X | Z$

(an assumption termed "conditional ignorability" by Rosenbaum and Rubin (1983),) then the causal effect $P(y|do(x)) = P^*(Y_x = y)$ can readily be evaluated to yield

$$P*(Y_x=y) = \sum_z P*(Y_x=y|z)P(z) \quad (36)$$

$$= \sum_z P*(Y_x=y|x,z)P(z) \quad \text{(using (35))}$$

$$= \sum_z P*(Y=y|x,z)P(z) \quad \text{(using (32))}$$

$$= \sum_z P(y|x,z)P(z).$$

The last expression contains no counterfactual quantities (thus permitting us to drop the asterisk from $P*$) and coincides precisely with the standard covariate-adjustment formula of Eq. (25).

We see that the assumption of conditional ignorability (35) qualifies Z as an admissible covariate for adjustment; it mirrors therefore the "back-door" criterion of Definition 3, which bases the admissibility of Z on an explicit causal structure encoded in the diagram.

The derivation above may explain why the potential-outcome approach appeals to mathematical statisticians; instead of constructing new vocabulary (e.g., arrows), new operators ($do(x)$) and new logic for causal analysis, almost all mathematical operations in this framework are conducted within the safe confines of probability calculus. Save for an occasional application of rule (34) or (32)), the analyst may forget that Y_x stands for a counterfactual quantity—it is treated as any other random variable, and the entire derivation follows the course of routine probability exercises.

This orthodoxy exacts a high cost: Instead of bringing the theory to the problem, the problem must be reformulated to fit the theory; all background knowledge pertaining to a given problem must first be translated into the language of counterfactuals (e.g., ignorability conditions) before analysis can commence. This translation may in fact be the hardest part of the problem. The reader may appreciate this aspect by attempting to judge whether the assumption of conditional ignorability (35), the key to the derivation of (36), holds in any familiar situation, say in the experimental setup of Fig. 2(a). This assumption reads: "the value that Y would obtain had X been x, is independent of X, given Z". Even the most experienced potential-outcome expert would be unable to discern whether any subset Z of covariates in Fig. 4 would satisfy this conditional independence condition.15 Likewise, to derive Eq. (35) in the language of potential-outcome (see (Pearl, 2000a, p. 223)), one would need to convey the structure of the chain $X \rightarrow W_3 \rightarrow Y$ using the cryptic expression: $W_{3x} \perp\!\!\!\perp \{Y_{w_3}, X\}$, read: "the value that W_3 would obtain had X been x is independent of the value that Y would obtain had W_3 been w_3 jointly with the value of X." Such assumptions are cast in a language so far removed from ordinary understanding of scientific theories that, for all practical purposes, they cannot be comprehended or ascertained by ordinary mortals. As a result, researchers in the graph-less potential-outcome camp rarely use "conditional ignorability" (35) to guide the choice of covariates; they view this condition as a hoped-for miracle of nature rather than a target to be achieved by reasoned design.16

Replacing "ignorability" with a conceptually meaningful condition (i.e., back-door) in a graphical model permits researchers to understand what conditions covariates must fulfill before they eliminate bias, what to watch for and what to think about when covariates are selected, and what experiments we can do to test, at least partially, if we have the knowledge needed for covariate selection.

Aside from offering no guidance in covariate selection, formulating a problem in the potential-outcome language encounters three additional hurdles. When counterfactual variables are not viewed as byproducts of a deeper, process-based model, it is hard to ascertain whether *all* relevant judgments have been articulated, whether the judgments articulated are *redundant*, or whether those judgments are *self-consistent.* The need to express, defend, and manage formidable counterfactual relationships of this type explain the slow acceptance of causal analysis among health scientists and statisticians, and why most economists and social scientists continue to use structural equation models (Wooldridge, 2002, Stock and Watson, 2003, Heckman, 2008) instead of the potential-outcome alternatives advocated in Angrist, Imbens, and Rubin (1996), Holland (1988), Sobel (1998, 2008).

On the other hand, the algebraic machinery offered by the counterfactual notation, $Y_x(u)$, once a problem is properly formalized, can be extremely powerful in refining assumptions (Angrist et al., 1996, Heckman and Vytlacil, 2005), deriving consistent estimands (Robins, 1986), bounding probabilities of necessary and sufficient causation (Tian and Pearl, 2000), and combining data from experimental and nonexperimental studies (Pearl, 2000a). The next subsection (5.3) presents a way of combining the best features of the two approaches. It is based on encoding causal assumptions in the language of diagrams, translating these assumptions into counterfactual notation, performing the mathematics in the algebraic language of counterfactuals (using (32), (33), and (34)) and, finally, interpreting the result in graphical terms or plain causal language. The mediation problem of Section 6.1 illustrates how such symbiosis clarifies the definition and identification of direct and indirect effects,[17] and how it overcomes difficulties that were deemed insurmountable in the exclusivist potential-outcome framework (Rubin, 2004, 2005).

5.3. Combining graphs and potential outcomes

The formulation of causal assumptions using graphs was discussed in Section 3. In this subsection we will systematize the translation of these assumptions from graphs to counterfactual notation.

Structural equation models embody causal information in both the equations and the probability function $P(u)$ assigned to the exogenous variables; the former is encoded as missing arrows in the diagrams the latter as missing (double arrows) dashed arcs. Each parent-child family (PA_i, X_i) in a causal diagram G corresponds to an equation in the model M. Hence, missing arrows encode exclusion assumptions, that is, claims that manipulating variables that are excluded from an equation will not change the outcome of the hypothetical experiment described by that equation. Missing dashed arcs encode independencies among error terms in two or more equations. For example, the absence of dashed arcs between a node Y and a set of nodes $\{Z_1, \ldots, Z_k\}$ implies that the corresponding background variables, U_Y and $\{U_{Z1}, \ldots, U_{Zk}\}$, are independent in $P(u)$.

These assumptions can be translated into the potential-outcome notation using two simple rules (Pearl, 2000a, p. 232); the first interprets the missing arrows in the graph, the second, the missing dashed arcs.

Exclusion restrictions: For every variable Y having parents PA_Y and for every set of endogenous variables S disjoint of PA_Y, we have (37) $Y_{pa_Y} = Y_{pa_Y,s}$.

Independence restrictions: If $Z_1, \ldots, Z_k$ is any set of nodes not connected to Y via dashed arcs, and $PA_1, \ldots, PA_k$ their respective sets of parents, we have

(38) $Y_{pa_Y} \perp\!\!\!\perp \{Z_{1\,pa_1}, \ldots, Z_{k\,pa_k}\}$.

The exclusion restrictions expresses the fact that each parent set includes *all* direct causes of the child variable, hence, fixing the parents of Y, determines the value of Y uniquely, and intervention on any other set S of (endogenous) variables can no longer affect Y. The independence restriction translates the independence between U_Y and $\{U_{Z_1}, \ldots, U_{Z_k}\}$ into independence between the corresponding potential-outcome variables. This follows from the observation that, once we set their parents, the variables in $\{Y, Z_1, \ldots, Z_k\}$ stand in functional relationships to the U terms in their corresponding equations.

As an example, consider the model shown in Fig. 5, which serves as the canonical representation for the analysis of instrumental variables (Angrist et al., 1996, Balke and Pearl, 1997). This model displays the following parent sets:

(39) $PA_Z = \{\emptyset\}$, $PA_X = \{Z\}$, $PA_Y = \{X\}$.

Consequently, the exclusion restrictions translate into:

$$X_z = X_{yz}$$
$$Z_y = Z_{xy} = Z_x = Z$$
(40) $$Y_x = Y_{xz}$$

the absence of any dashed arc between Z and $\{Y, X\}$ translates into the independence restriction

(41) $Z \perp\!\!\!\perp \{Y_x, X_z\}$.

This is precisely the condition of randomization; Z is independent of all its non-descendants, namely independent of U_X and U_Y which are the exogenous parents of Y and X, respectively. (Recall that the exogenous parents of any variable, say Y, may be replaced by the counterfactual variable Y_{pa_Y}, because holding PA_Y constant renders Y a deterministic function of its exogenous parent

U_Y.)

The role of graphs is not ended with the formulation of causal assumptions. Throughout an algebraic derivation, like the one shown in Eq. (36), the analyst may need to employ additional assumptions that are entailed by the original exclusion and independence assumptions, yet are not shown explicitly in their respective algebraic expressions. For example, it is hardly straightforward to show that the assumptions of Eqs. (40)–(41) imply the conditional independence $(Y_x \perp\!\!\!\perp Z | \{X_z, X\})$ but do not imply the conditional independence $(Y_x \perp\!\!\!\perp Z | X)$. These are not easily derived by algebraic means alone. Such implications can, however, easily be tested in the graph of Fig. 5 using the graphical reading for conditional independence (Definition 1). (See (Pearl, 2000a, pp. 16–17, 213–215).) Thus, when the need arises to employ independencies in the course of a derivation, the graph may assist the procedure by vividly displaying the independencies that logically follow from our assumptions.

6. Counterfactuals at Work
6.1. Mediation: Direct and indirect effects
6.1.1. Direct versus total effects

The causal effect we have analyzed so far, $P(y|do(x))$, measures the *total* effect of a variable (or a set of variables) *X* on a response variable *Y*. In many cases, this quantity does not adequately represent the target of investigation and attention is focused instead on the *direct* effect of *X* on *Y*. The term "direct effect" is meant to quantify an effect that is not mediated by other variables in the model or, more accurately, the sensitivity of *Y* to changes in *X* while all other factors in the analysis are held fixed. Naturally, holding those factors fixed would sever all causal paths from *X* to *Y* with the exception of the direct link *X* → *Y*, which is not intercepted by any intermediaries.

A classical example of the ubiquity of direct effects involves legal disputes over race or sex discrimination in hiring. Here, neither the effect of sex or race on applicants' qualification nor the effect of qualification on hiring are targets of litigation. Rather, defendants must prove that sex and race do not *directly* influence hiring decisions, whatever indirect effects they might have on hiring by way of applicant qualification.

From a policy making viewpoint, an investigator may be interested in decomposing effects to quantify the extent to which racial salary disparity is due to educational disparity, or, taking a health-care example, the extent to which sensitivity to a given exposure can be reduced by eliminating sensitivity to an intermediate factor, standing between exposure and outcome. Another example concerns the identification of neural pathways in the brain or the structural features of protein-signaling networks in molecular biology (Brent and Lok, 2005). Here, the decomposition of effects into their direct and indirect components carries theoretical scientific importance, for it tells us "how nature works" and, therefore, enables us to predict behavior under a rich variety of conditions.

Yet despite its ubiquity, the analysis of mediation has long been a thorny issue in the social and behavioral sciences (Judd and Kenny, 1981, Baron and Kenny, 1986, Muller, Judd, and Yzerbyt, 2005, Shrout and Bolger, 2002, MacKinnon, Fairchild, and Fritz, 2007a) primarily because structural equation modeling in those sciences were deeply entrenched in linear analysis, where the distinction between causal parameters and their regressional interpretations can easily be conflated.[18] As demands grew to tackle problems involving binary and categorical variables, researchers could no longer define direct and indirect effects in terms of structural or regressional coefficients, and all attempts to extend the linear paradigms of effect decomposition to non-linear systems produced distorted results (MacKinnon, Lockwood, Brown, Wang, and Hoffman, 2007b). These difficulties have accentuated the need to redefine and derive causal effects from first principles, uncommitted to distributional assumptions or a particular parametric form of the equations. The structural methodology presented in this paper adheres to this philosophy and it has produced indeed a principled solution to the mediation problem, based on the counterfactual reading of structural equations (29). The following subsections summarize the method and its solution.

6.1.2. Controlled direct-effects

A major impediment to progress in mediation analysis has been the lack of notational facility for expressing the key notion of "holding the mediating variables fixed" in the definition of direct effect. Clearly, this notion must be interpreted as (hypothetically) setting the intermediate variables to constants by physical intervention, not by analytical means such as selection, regression, conditioning, matching or adjustment. For example, consider the simple mediation models of Fig. 6, where the error terms (not shown explicitly) are assumed to be independent. It will not be sufficient to measure the association between gender (X) and hiring (Y) for a given level of qualification (Z), (see Fig. 6(b)) because, by conditioning on the mediator Z, we create spurious associations between X and Y through W_2, even when there is no direct effect of X on Y (Pearl, 1998, Cole and Hernán, 2002).

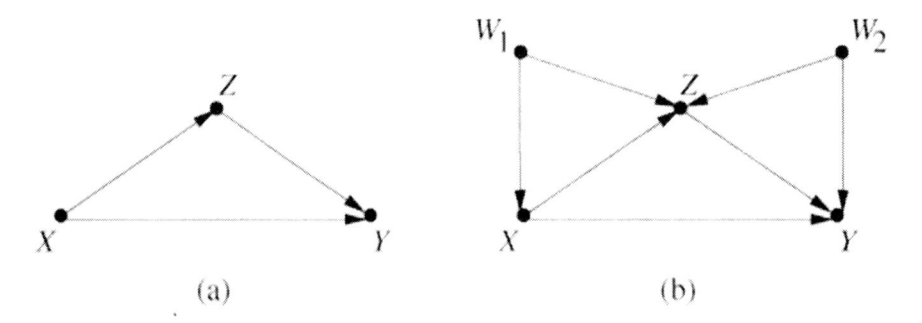

Figure 6: (a) A generic model depicting mediation through Z with no confounders, and (b) with two confounders, W_1 and W_2.

Using the *do(x)* notation, enables us to correctly express the notion of "holding Z fixed" and obtain a simple definition of the *controlled direct effect* of the transition from $X = x$ to $X = x'$:

$$CDE \triangleq E(Y|do(x), do(z)) - E(Y|do(x'), do(z))$$

or, equivalently, using counterfactual notation:

$$CDE \triangleq E(Y_{xz}) - E(Y_{x'z})$$

where Z is the set of all mediating variables. The readers can easily verify that, in linear systems, the controlled direct effect reduces to the path coefficient of the link $X \rightarrow Y$ (see footnote 12) regardless of whether confounders are present (as in Fig. 6(b)) and regardless of whether the error terms are correlated or not.

This separates the task of definition from that of identification, as demanded by Section 4.1. The identification of *CDE* would depend, of course, on whether confounders are present and whether they can be neutralized by adjustment, but these do not alter its definition. Nor should trepidation about infeasibility of the action *do(gender = male)* enter the definitional phase of the study, Definitions apply to symbolic models, not to human biology. Graphical identification conditions for expressions of the type $E(Y|do(x), do(z_1), do(z_2), \ldots, do(z_k))$ in the presence of unmeasured confounders were derived by Pearl and Robins (1995) (see Pearl (2000a, Chapter 4) and invoke sequential application of the back-door conditions discussed in Section 3.2.

6.1.3. Natural direct effects

In linear systems, the direct effect is fully specified by the path coefficient attached to the link from X to Y; therefore, the direct effect is independent of the values at which we hold Z. In nonlinear systems, those values would, in general, modify the effect of X on Y and thus should be chosen carefully to represent the target policy under analysis. For example, it is not uncommon to find employers who prefer males for the high-paying jobs (i.e., high z) and females for low-paying jobs (low z).

When the direct effect is sensitive to the levels at which we hold Z, it is often more meaningful to define the direct effect relative to some "natural" base-line level that may vary from individual to individual, and represents the level of Z just before the change in X. Conceptually, we can define the natural direct effect $DE_{x,x'}(Y)$ as the expected change in Y induced by changing X from x to x' while keeping all mediating factors constant at whatever value they *would have obtained* under *do(x)*. This hypothetical change, which Robins and Greenland (1992) conceived and called "pure" and Pearl (2001) formalized and analyzed under the rubric "natural," mirrors what lawmakers instruct us to consider in race or sex discrimination cases: "The central question in any employment-discrimination case is whether the employer would have taken the same action had the employee been of a different race (age, sex, religion, national origin etc.) and everything else had been the same." (In *Carson versus Bethlehem Steel Corp.*, 70 FEP Cases 921, 7th Cir. (1996)).

Extending the subscript notation to express nested counterfactuals, Pearl (2001) gave a formal definition for the "natural direct effect":

(42) $DE_{x,x'}(Y) = E(Y_{x',Z_x}) - E(Y_x)$.

Here, Y_{x',Z_x} represents the value that Y would attain under the operation of setting X to x' and, simultaneously, setting Z to whatever value it would have obtained under the setting $X = x$. We see that $DE_{x,x'}(Y)$, the natural direct effect of the transition from x to x', involves probabilities of *nested counterfactuals* and cannot be written in terms of the *do(x)* operator. Therefore, the natural direct effect cannot in general be identified, even with the help of ideal, controlled experiments (see footnote 8 for intuitive explanation). However, aided by the surgical definition of Eq. (29) and the notational power of nested counterfactuals, Pearl (2001) was nevertheless able to show that, if certain assumptions of "no confounding" are deemed valid, the natural direct effect can be reduced to

(43)
$$DE_{x,x'}(Y) = \sum_z [E(Y|do(x',z)) - E(Y|do(x,z))]P(z|do(x)).$$

The intuition is simple; the natural direct effect is the weighted average of the controlled direct effect, using the causal effect $P(z|do(x))$ as a weighing function.

One condition for the validity of (43) is that $Z_x \perp\!\!\!\perp Y_{x',z} | W$ holds for some set W of measured covariates. This technical condition in itself, like the ignorability condition of (35), is close to meaningless for most investigators, as it is not phrased in terms of realized variables. The surgical interpretation of counterfactuals (29) can be invoked at this point to unveil the graphical interpretation of this condition. It states that W should be admissible (i.e., satisfy the back-door condition) relative the path(s) from Z to Y. This condition, satisfied by W_2 in Fig. 6(b), is readily comprehended by empirical researchers, and the task of selecting such measurements, W, can then be guided by the available scientific knowledge. Additional graphical and counterfactual conditions for identification are derived in Pearl (2001) Petersen et al. (2006) and Imai, Keele, and Yamamoto (2008).

In particular, it can be shown (Pearl, 2001) that expression (43) is both valid and identifiable in Markovian models (i.e., no unobserved confounders) where each term on the right can be reduced to a "*do*-free" expression using Eq. (24) or (25) and then estimated by regression.

For example, for the model in Fig. 6(b), Eq. (43) reads:

(44)
$$DE_{x,x'}(Y) = \sum_{z} \sum_{w_1} P(w_1)[E(Y|x',z,w_1)) - E(Y|x,z,w_1))] \sum_{w_2} P(z|x,w_2)P(w_2).$$

while for the confounding-free model of Fig. 6(a) we have:

(45)
$$DE_{x,x'}(Y) = \sum_{z} [E(Y|x',z) - E(Y|x,z)]P(z|x).$$

Both (44) and (45) can easily be estimated by a two-step regression.

6.1.4. Natural indirect effects

Remarkably, the definition of the natural direct effect (42) can be turned around and provide an operational definition for the *indirect effect* — a concept shrouded in mystery and controversy, because it is impossible, using the *do(x)* operator, to disable the direct link from X to Y so as to let X influence Y solely via indirect paths.

The *natural indirect effect*, *IE*, of the transition from x to x' is defined as the expected change in Y affected by holding X constant, at X = x, and changing Z to whatever value it would have attained had X been set to X = x'. Formally, this reads (Pearl, 2001):

(46) $IE_{x,x'}(Y) \triangleq E[(Y_{x,Z_{x'}}) - E(Y_x)]$,

which is almost identical to the direct effect (Eq. (42)) save for exchanging x and x' in the first term.

Indeed, it can be shown that, in general, the total effect TE of a transition is equal to the *difference* between the direct effect of that transition and the indirect effect of the reverse transition. Formally,

(47) $TE_{x,x'}(Y) \triangleq E(Y_{x'} - Y_x) = DE_{x,x'}(Y) - IE_{x',x}(Y)$.

In linear systems, where reversal of transitions amounts to negating the signs of their effects, we have the standard additive formula

(48) $TE_{x,x'}(Y) = DE_{x,x'}(Y) + IE_{x,x'}(Y)$.

Since each term above is based on an independent operational definition, this equality constitutes a formal justification for the additive formula used routinely in linear systems.

Note that, although it cannot be expressed in *do*-notation, the indirect effect has clear policy-making implications. For example: in the hiring discrimination context, a policy maker may be interested in predicting the gender mix in the work force if gender bias is eliminated and all applicants are treated equally—say, the same way that males are currently treated. This quantity will be given by the indirect effect of gender on hiring, mediated by factors such as education and aptitude, which may be gender-dependent.

More generally, a policy maker may be interested in the effect of issuing a directive to a select set of subordinate employees, or in carefully controlling the routing of messages in a network of interacting agents. Such applications motivate the analysis of *path-specific effects*, that is, the effect of X on Y through a selected set of paths (Avin, Shpitser, and Pearl, 2005).

In all these cases, the policy intervention invokes the selection of signals to be sensed, rather than variables to be fixed. Pearl (2001) has suggested therefore that *signal sensing* is more fundamental to the notion of causation than *manipulation*; the latter being but a crude way of stimulating the former in experimental setup. The mantra "No causation without manipulation" must be rejected. (See (Pearl, 2009b, Section 11.4.5).)

It is remarkable that counterfactual quantities like *DE* and *IE* that could not be expressed in terms of *do(x)* operators, and appear therefore void of empirical content, can, under certain conditions be estimated from empirical studies, and serve to guide policies. Awareness of this potential should embolden researchers to go through the definitional step of the study and freely articulate the target quantity *Q(M)* in the language of science, i.e., counterfactuals, despite the seemingly speculative nature of each assumption in the model (Pearl, 2000b).

6.2. The Mediation Formula: a simple solution to a thorny problem

This subsection demonstrates how the solution provided in equations (45) and (48) can be applied to practical problems of assessing mediation effects in non-linear models. We will use the simple mediation model of Fig. 6(a), where all error terms (not shown explicitly) are assumed to be mutually independent, with the understanding that adjustment for appropriate sets of covariates *W* may be necessary to achieve this independence and that integrals should replace summations when dealing with continuous variables (Imai et al., 2008).

Combining (45) and (48), the expression for the indirect effect, *IE*, becomes:

(49)
$$IE_{x,x'}(Y) = \sum_{z} E(Y|x,z)[P(z|x') - P(z|x)]$$

which provides a general formula for mediation effects, applicable to any nonlinear system, any distribution (of *U*), and any type of variables. Moreover, the formula is readily estimable by regression. Owed to its generality and ubiquity, I will refer to this expression as the "Mediation Formula."

The Mediation Formula represents the average increase in the outcome *Y* that the transition from $X = x$ to $X = x'$ is expected to produce absent any direct effect of *X* on *Y*. Though based on solid causal principles, it embodies no causal assumption other than the generic mediation structure of Fig. 6(a). When the outcome *Y* is binary (e.g., recovery, or hiring) the ratio (1 − *IE/TE*) represents the fraction of responding individuals who owe their response to direct paths, while (1 − *DE/TE*) represents the fraction who owe their response to *Z*-mediated paths.

The Mediation Formula tells us that *IE* depends only on the expectation of the counterfactual Y_{xz}, not on its functional form $f_Y(x, z, u_Y)$ or its distribution $P(Y_{xz} = y)$. It calls therefore for a two-step regression which, in principle, can be performed non-parametrically. In the first step we regress *Y* on *X* and *Z*, and obtain the estimate

$g(x, z) = E(Y|x, z)$

for every (*x*, *z*) cell. In the second step we estimate the expectation of $g(x, z)$ conditional on $X = x'$ and $X = x$, respectively, and take the difference:

$IE_{x,x'}(Y) = E_z(g(x, z)|x') - E_z(g(x, z)|x)$

Nonparametric estimation is not always practical. When *Z* consists of a vector of several mediators, the dimensionality of the problem would prohibit the estimation of $E(Y|x, z)$ for every (*x*, *z*) cell, and the need arises to use parametric approximation. We can then choose any convenient parametric form for $E(Y|x, z)$ (e.g., linear, logit, probit), estimate the parameters separately (e.g., by regression or maximum likelihood methods), insert the parametric approximation into (49) and estimate its two conditional expectations (over *z*) to get the mediated effect (VanderWeele, 2009).

Let us examine what the Mediation Formula yields when applied to both linear and non-linear versions of model 6(a). In the linear case, the structural model reads:

$$x = u_X$$
$$z = b_x x + u_Z$$
$$y = c_x x + c_z z + u_Y \quad (50)$$

Computing the conditional expectation in (49) gives

$E(Y|x, z) = E(c_x x + c_z z + u_Y) = c_x x + c_z z$

and yields

(51)
$$IE_{x,x'}(Y)=\sum_z(c_x x+c_z z)[P(z|x') - P(z|x)].$$
$$=c_z[E(Z|x') - E(Z|x)]$$

(52) $= (x' - x)(c_z b_x)$

(53) $= (x' - x)(b - c_x)$

where b is the total effect coefficient, $b = (E(Y|x') - E(Y|x))/(x' - x) = c_x + c_z b_x$.

We thus obtained the standard expressions for indirect effects in linear systems, which can be estimated either as a difference in two regression coefficients (Eq. 53) or a product of two regression coefficients (Eq. 52), with Y regressed on both X and Z. (see (MacKinnon et al., 2007b)). These two strategies do not generalize to non-linear system as we shall see next.

Suppose we apply (49) to a non-linear process (Fig. 7) in which X, Y, and Z are binary variables, and Y and Z are given by the Boolean formula

$$Y = \text{AND}(x, e_x) \vee \text{AND}(z, e_z) \quad x, z, e_x, e_z = 0, 1$$
$$z = \text{AND}(x, e_{xz}) \quad\quad\quad\quad\quad\quad z, e_{xz} = 0, 1$$

Such disjunctive interaction would describe, for example, a disease Y that would be triggered either by X directly, if enabled by e_x, or by Z, if enabled by e_z. Let us further assume that e_x, e_z and e_{xz} are three independent Bernoulli variables with probabilities p_x, p_z, and p_{xz}, respectively.

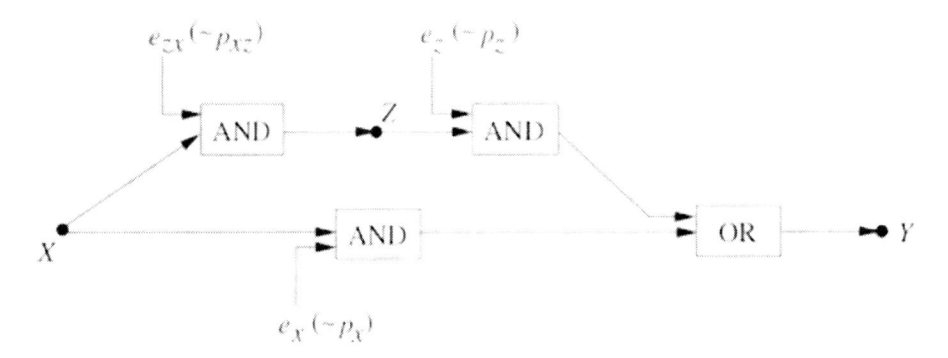

Figure 7: Stochastic non-linear model of mediation. All variables are binary.

As investigators, we are not aware, of course, of these underlying mechanisms; all we know is that X, Y, and Z are binary, that Z is hypothesized to be a mediator, and that the assumption of nonconfoundedness permits us to use the Mediation Formula (49) for estimating the Z-mediated effect of X on Y. Assume that our plan is to conduct a nonparametric estimation of the terms in (49) over a very large sample drawn from P(x, y.z); it is interesting to ask what the asymptotic value of the Mediation Formula would be, as a function of the model parameters: p_x, p_z, and p_{xz}.

From knowledge of the underlying mechanism, we have:

$$P(Z=1|x) = p_{xz} x \qquad x=0,1$$
$$P(Y=1|x,z) = p_x x + p_z z - p_x p_z xz \qquad x,z=0,1$$

Therefore,

$$E(Z|x) = p_{xz} x \qquad\qquad x = 0, 1$$
$$E(Y|x, z) = x p_x + z p_z - x z p_x p_z \qquad x, z = 0, 1$$
$$E(Y|x) = \sum_z E(Y|x, z) P(z|x)$$
$$= x p_x + (p_z - x p_x p_z) E(Z|x)$$
$$= x(p_x + p_{xz} p_z - x p_x p_z p_{xz}) \qquad x = 0, 1$$

Taking $x = 0$, $x' = 1$ and substituting these expressions in (45), (48), and (49) yields

(54) $IE(Y) = p_z p_{xz}$

(55) $DE(Y) = p_x$

(56) $TE(Y) = p_z p_{xz} + p_x + p_x p_z p_{xz}$

Two observations are worth noting. First, we see that, despite the non-linear interaction between the two causal paths, the parameters of one do not influence on the causal effect mediated by the other. Second, the total effect is not the sum of the direct and indirect effects. Instead, we have:

$TE = DE + IE - DE \cdot IE$

which means that a fraction $DE \cdot IE/TE$ of outcome cases triggered by the transition from $X = 0$ to $X = 1$ are triggered simultaneously, through both causal paths, and would have been triggered even if one of the paths was disabled.

Now assume that we choose to approximate $E(Y|x, z)$ by the linear expression

(57) $g(x, z) = a_0 + a_1 x + a_2 z$.

After fitting the a's parameters to the data (e.g., by OLS) and substituting in (49) one would obtain

(58)

$$IE_{x,x'}(Y) = \sum_z (a_0 + a_1 x + a_2 z)[P(z|x') - P(z|x)]$$
$$= a_2[E(Z|x') - E(Z|x)]$$

which holds whenever we use the approximation in (57), regardless of the underlying mechanism.

If the correct data-generating process was the linear model of (50), we would obtain the expected estimates $a_2 = c_z$, $E(z|x') - E(z|x') = b_x(x' - x)$ and

$IE_{x,x'}(Y) = b_x c_z(x' - x)$.

If however we were to apply the approximation in (57) to data generated by the nonlinear model of Fig. 7, a distorted solution would ensue; a_2 would evaluate to

$$a_2 = \sum_x [E(Y|x, z=1) - E(Y|x, z=0)]P(x)$$
$$= P(x=1)[E(Y|x=1, z=1) - E(Y|x=1, z=0)]$$
$$= P(x=1)[(p_x + p_z - p_x p_z) - p_x]$$
$$= P(x=1) p_z (1 - p_x),$$

$E(z|x') - E(z|x)$ would evaluate to $p_{xz}(x' - x)$, and (58) would yield the approximation

$$\widehat{IE}_{x,x'}(Y) = a_2[E(Z|x') - E(Z|x)]$$
$$= p_{xz} P(x=1) p_z (1 - p_x)$$

(59)

We see immediately that the result differs from the correct value $p_z p_{xz}$ derived in (54). Whereas the approximate value depends on $P(x = 1)$, the correct value shows no such dependence, and rightly so; no causal effect should depend on the probability of the causal variable.

Fortunately, the analysis permits us to examine under what condition the distortion would be significant. Comparing (59) and (54) reveals that the approximate method always underestimates the indirect effect and the distortion is minimal for high values of $P(x = 1)$ and $(1- p_x)$.

Had we chosen to include an interaction term in the approximation of $E(Y|x, z)$, the correct result would obtain. To witness, writing

$E(Y|x, z) = a_0 + a_1 x + a_2 z + a_3 xz$,

a_2 would evaluate to p_z, a_3 to $p_x p_z$, and the correct result obtains through:

$$IE_{x,x'}(Y) = \sum_z (a_0 + a_1 x + a_2 z + a_3 xz)[P(z|x') - P(z|x)]$$
$$= (a_2 + a_3 x)[E(Z|x') - E(Z|x)]$$
$$= (a_2 + a_3 x) p_{xz}(x' - x)$$
$$= (p_z - p_x p_z x) p_{xz}(x' - x)$$

We see that, in addition to providing causally-sound estimates for mediation effects, the Mediation Formula also enables researchers to evaluate analytically the effectiveness of various parametric specifications relative to any assumed model. This type of analytical "sensitivity analysis" has been used extensively in statistics for parameter estimation, but could not be applied to mediation analysis, owed to the absence of an objective target quantity that captures the notion of indirect effect in both linear and non-linear systems, free of parametric assumptions. The Mediation Formula of Eq. (49) explicates this target quantity formally, and casts it in terms of estimable quantities.

The derivation of the Mediation Formula was facilitated by taking seriously the four steps of the structural methodology (Section 4) together with the graphical-counterfactual-structural symbiosis spawned by the surgical interpretation of counterfactuals (Eq. (29)).

In contrast, when the mediation problem is approached from an exclusivist potential-outcome viewpoint, void of the structural guidance of Eq. (29), counterintuitive definitions ensue, carrying the label "principal stratification" (Rubin, 2004, 2005), which are at variance with common understanding of direct and indirect effects. For example, the direct effect is definable only in units absent of indirect effects. This means that a grandfather would be deemed to have no direct effect on his grandson's behavior in families where he has had some effect on the father. This precludes from the analysis all typical families, in which a father and a grandfather have simultaneous, complementary influences on children's upbringing. In linear systems, to take an even sharper example, the direct effect would be undefined whenever indirect paths exist from the cause to its effect. The emergence of such paradoxical conclusions underscores the wisdom, if not necessity of a symbiotic analysis, in which the counterfactual notation $Y_x(u)$ is governed by its structural definition, Eq. (29).19

6.3. Causes of effects and probabilities of causation

The likelihood that one event *was the cause* of another guides much of what we understand about the world (and how we act in it). For example, knowing whether it was the aspirin that cured my headache or the TV program I was watching would surely affect my future use of aspirin. Likewise, to take an example from common judicial standard, judgment in favor of a plaintiff should be made if and only if it is "more probable than not" that the damage would not have occurred *but for* the defendant's action (Robertson, 1997).

These two examples fall under the category of "causes of effects" because they concern situations in which we observe both the effect, $Y = y$, and the putative cause $X = x$ and we are asked to assess, counterfactually, whether the former would have occurred absent the latter.

We have remarked earlier (footnote 8) that counterfactual probabilities conditioned on the outcome cannot in general be identified from observational or even experimental studies. This does not mean however that such probabilities are useless or void of empirical content; the structural perspective may guide us in fact toward discovering the conditions under which they can be assessed from data, thus defining the empirical content of these counterfactuals.

Following the 4-step process of structural methodology – define, assume, identify, and estimate – our first step is to express the target quantity in counterfactual notation and verify that it is well defined, namely, that it can be computed unambiguously from any fully-specified causal model.

In our case, this step is simple. Assuming binary events, with $X = x$ and $Y = y$ representing treatment and outcome, respectively, and $X = x'$, $Y = y'$ their negations, our target quantity can be formulated directly from the English sentence:

"Find the probability that Y would be y' had X been x', given that, in reality, Y is actually y and X is x,"

to give:

(60) $PN(x, y) = P(Y_{x'}' = y' | X = x, Y = y)$

This counterfactual quantity, which Robins and Greenland (1989b) named "probability of causation" and Pearl (2000a, p. 296) named "probability of necessity" (PN), to be distinguished from two other nuances of "causation," is certainly computable from any fully specified structural model, i.e., one in which $P(u)$ and all functional relationships are given. This follows from the fact that every structural model defines a joint distribution of counterfactuals, through Eq. (29).

Having written a formal expression for PN, Eq. (60), we can move on to the formulation and identification phases and ask what assumptions would permit us to identify PN from empirical studies, be they observational, experimental or a combination thereof.

This problem was analyzed in Pearl (2000a, Chapter 9) and yielded the following results:

Theorem 4 *If Y is monotonic relative to X, i.e., $Y_1(u) \geq Y_0(u)$, then PN is identifiable whenever the causal effect $P(y|do(x))$ is identifiable and, moreover,*

(61)
$$PN = \frac{P(y|x) - P(y|x')}{P(y|x)} + \frac{P(y|x') - P(y|do(x'))}{P(x, y)}.$$

The first term on the r.h.s. of (61) is the familiar excess risk ratio (ERR) that epidemiologists have been using as a surrogate for PN in court cases (Cole, 1997, Robins and Greenland, 1989b). The second term represents the *correction* needed to account for confounding bias, that is, $P(y|do(x')) \neq P(y|x')$.

This suggests that monotonicity and unconfoundedness were tacitly assumed by the many authors who proposed or derived ERR as a measure for the "fraction of exposed cases that are attributable to the exposure" (Greenland, 1999).

Equation (61) thus provides a more refined measure of causation, which can be used in situations where the causal effect $P(y|do(x))$ can be estimated from either randomized trials or graph-assisted observational studies (e.g., through Theorem 3 or Eq. (25)). It can also be shown (Tian and Pearl, 2000) that the expression in (61) provides a lower bound for PN in the general, nonmonotonic case. (See also (Robins and Greenland, 1989a).) In particular, the tight upper and lower bounds on PN are given by:

(62)
$$\max\left\{0, \frac{P(y) - P(y|do(x'))}{P(x,y)}\right\} \leq PN \leq \min\left\{1, \frac{P(y'|do(x')) - P(x',y')}{P(x,y)}\right\}$$

It is worth noting that, in drug related litigation, it is not uncommon to obtain data from both experimental and observational studies. The former is usually available at the manufacturer or the agency that approved the drug for distribution (e.g., FDA), while the latter is easy to obtain by random surveys of the population. In such cases, the standard lower bound used by epidemiologists to establish legal responsibility, the Excess Risk Ratio, can be improved substantially using the corrective term of Eq. (61). Likewise, the upper bound of Eq. (62) can be used to exonerate drug-makers from legal responsibility. Cai and Kuroki (2006) analyzed the statistical properties of PN.

Pearl (2000a, p. 302) shows that combining data from experimental and observational studies which, taken separately, may indicate no causal relations between X and Y, can nevertheless bring the lower bound of Eq. (62) to unity, thus implying causation *with probability one*.

Such extreme results dispel all fears and trepidations concerning the empirical content of counterfactuals (Dawid, 2000, Pearl, 2000b). They demonstrate that a quantity PN which at first glance appears to be hypothetical, ill-defined, untestable and, hence, unworthy of scientific analysis is nevertheless definable, testable and, in certain cases, even identifiable. Moreover, the fact that, under certain combination of data, and making no assumptions whatsoever, an important legal claim such as "the plaintiff would be alive had he not taken the drug" can be ascertained with probability approaching one, is a remarkable tribute to formal analysis.

Another counterfactual quantity that has been fully characterized recently is the Effect of Treatment on the Treated (ETT):

$ETT = P(Y_x = y | X = x')$

ETT has been used in econometrics to evaluate the effectiveness of social programs on their participants (Heckman, 1992) and has long been the target of research in epidemiology, where it came to be known as "the effect of exposure on the exposed," or "standardized morbidity" (Miettinen, 1974; Greenland and Robins, 1986).

Shpitser and Pearl (2009) have derived a complete characterization of those models in which ETT can be identified from either experimental or observational studies. They have shown that, despite its blatant counterfactual character, (e.g., "I just took an aspirin, perhaps I shouldn't have?") ETT can be evaluated from experimental studies in many, though not all cases. It can also be evaluated from observational studies whenever a sufficient set of covariates can be measured that satisfies the back-door criterion and, more generally, in a wide class of graphs that permit the identification of conditional interventions.

These results further illuminate the empirical content of counterfactuals and their essential role in causal analysis. They prove once again the triumph of logic and analysis over traditions that a-priori exclude from the analysis quantities that are not testable in isolation. Most of all, they demonstrate the effectiveness and viability of the *scientific* approach to causation whereby the dominant paradigm is to model the activities of Nature, rather than those of the experimenter. In contrast to the ruling paradigm of conservative statistics, we begin with relationships that we know in advance will never be estimated, tested or falsified. Only after assembling a host of such relationships and judging them to faithfully represent our theory about how Nature operates, we ask whether the parameter of interest, crisply defined in terms of those theoretical relationships, can be estimated consistently from empirical data and how. It often does, to the credit of progressive statistics.

7. Conclusions

Traditional statistics is strong in devising ways of describing data and inferring distributional parameters from sample. Causal inference requires two additional ingredients: a science-friendly language for articulating causal knowledge, and a mathematical machinery for processing that knowledge, combining it with data and drawing new causal conclusions about a phenomenon. This paper surveys recent advances in causal analysis from the unifying perspective of the structural theory of causation and shows how statistical methods can be supplemented with the needed ingredients. The theory invokes non-parametric structural equations models as a formal and meaningful language for defining causal quantities, formulating causal assumptions, testing identifiability, and explicating many concepts used in causal discourse. These include: randomization, intervention, direct and indirect effects, confounding, counterfactuals, and attribution. The algebraic component of the structural language coincides with the potential-outcome framework, and its graphical component embraces Wright's method of path diagrams. When unified and synthesized, the two components offer statistical investigators a powerful and comprehensive methodology for empirical research.

Footnotes

1. By "untested" I mean untested using frequency data in nonexperimental studies.

2. Clearly, $P(Y = y | do(X = x))$ is equivalent to $P(Y_x = y)$. This is what we normally assess in a controlled experiment, with X randomized, in which the distribution of Y is estimated for each level x of X.

3. Linear relations are used here for illustration purposes only; they do not represent typical disease-symptom relations but illustrate the historical development of path analysis. Additionally, we will use standardized variables, that is, zero mean and unit variance.

4. Additional implications called "dormant independence" (Shpitser and Pearl, 2008) may be deduced from some graphs with correlated errors (Verma and Pearl, 1990).

5. A simple proof of the Causal Markov Theorem is given in Pearl (2000a, p. 30). This theorem was first presented in Pearl and Verma (1991), but it is implicit in the works of Kiiveri, Speed, and Carlin (1984) and others. Corollary 1 was named "Manipulation Theorem" in Spirtes et al. (1993), and is also implicit in Robins' (1987) *G*-computation formula. See Lauritzen (2001).

6. For clarity, we drop the (superfluous) subscript 0 from x_0 and z_{20}.

7. Before applying this criterion, one may delete from the causal graph all nodes that are not ancestors of Y.

8. The reason for this fundamental limitation is that no death case can be tested twice, with and without treatment. For example, if we measure equal proportions of deaths in the treatment and control groups, we cannot tell how many death cases are actually attributable to the treatment itself; it is quite possible that many of those who died under treatment would be alive if untreated and, simultaneously, many of those who survived with treatment would have died if not treated.

9. Connections between structural equations and a restricted class of counterfactuals were first recognized by Simon and Rescher (1966). These were later generalized by Balke and Pearl (1995), using surgeries (Eq. (29)), thus permitting endogenous variables to serve as counterfactual antecedents. The term "surgery definition" was used in Pearl (2000a, Epilogue) and criticized by Cartwright (2007) and Heckman (2005), (see Pearl (2009b, pp. 362–3, 374–9 for

rebuttals)).

10. The distinction between general, or population-level causes (e.g., "Drinking hemlock causes death") and singular or unit-level causes (e.g., "Socrates' drinking hemlock caused his death"), which many philosophers have regarded as irreconcilable (Eells, 1991), introduces no tension at all in the structural theory. The two types of sentences differ merely in the level of situation-specific information that is brought to bear on a problem, that is, in the specificity of the evidence e that enters the quantity $P(Y_x = y|e)$. When e includes *all* factors u, we have a deterministic, unit-level causation on our hand; when e contains only a few known attributes (e.g., age, income, occupation etc.) while others are assigned probabilities, a population-level analysis ensues.

11. For example, "The probability is 80% that Joe belongs to the class of patients who will be cured if they take the drug and die otherwise."

12. Note that β in Eq. (1), the incremental causal effect of X on Y, is defined procedurally by

$$\beta \triangleq E(Y|do(x_0+1)) - E(Y|do(x_0))$$
$$= \frac{\partial}{\partial x} E(Y|do(x))$$
$$= \frac{\partial}{\partial x} E(Y_x).$$

Naturally, all attempts to give β statistical interpretation have ended in frustrations (Holland, 1988, Whittaker, 1990, Wermuth, 1992, Wermuth and Cox, 1993), some persisting well into the 21st century (Sobel, 2008).   13. These equivalence classes were later called "principal stratification" by Frangakis and Rubin (2002). Looser bounds were derived earlier by Robins (1989) and Manski (1990).

14. The notation $Y \perp\!\!\!\perp X | Z$ stands for the conditional independence relationship $P(Y = y, X = x | Z = z) = P(Y = y | Z = z) P(X = x | Z = z)$ (Dawid, 1979).

15. Inquisitive readers are invited to guess whether $X_2 \perp\!\!\!\perp Z | Y$ holds in Fig. 2(a), then reflect on why causality is so slow in penetrating statistical education.

16. The opaqueness of counterfactual independencies explains why many researchers within the potential-outcome camp are unaware of the fact that adding a covariate to the analysis (e.g., Z_3 in Fig. 4, Z in Fig. 5 may actually *increase* confounding bias in propensity-score matching. Paul Rosenbaum, for example, writes: "there is little or no reason to avoid adjustment for a true covariate, a variable describing subjects before treatment" (Rosenbaum, 2002, p. 76). Rubin (2009) goes as far as stating that refraining from conditioning on an available measurement is "nonscientific ad hockery" for it goes against the tenets of Bayesian philosophy (see (Pearl, 2009c,d, Heckman and Navarro-Lozano, 2004) for a discussion of this fallacy).

17. Such symbiosis is now standard in epidemiology research (Robins, 2001, Petersen, Sinisi, and van der Laan, 2006, VanderWeele and Robins, 2007, Hafeman and Schwartz, 2009, VanderWeele, 2009) yet still lacking in econometrics (Heckman, 2008, Imbens and Wooldridge, 2009).

18. All articles cited above define the direct and indirect effects through their regressional interpretations; I am not aware of any article in this tradition that formally adapts a causal interpretation, free of estimation-specific parameterization.

19. Such symbiosis is now standard in epidemiology research (Robins, 2001, Petersen et al., 2006, VanderWeele and Robins, 2007, Hafeman and Schwartz, 2009, VanderWeele, 2009) and is making its way slowly toward the social and behavioral sciences.

*. Portions of this paper are adapted from Pearl (2000a, 2009a,b); I am indebted to Elja Arjas, Sander Greenland, David MacKinnon, Patrick Shrout, and many readers of the UCLA Causality Blog (http://www.mii.ucla.edu/causality/) for reading and commenting on various segments of this manuscript, and

especially to Erica Moodie and David Stephens for their thorough editorial input. This research was supported in parts by NIH grant #1R01 LM009961-01, NSF grant #IIS-0914211, and ONR grant #N000-14-09-1-0665.

References

Angrist J, Imbens G, Rubin D, authors. 1996;"Identification of causal effects using instrumental variables (with comments),". Journal of the American Statistical Association. 91:444–472. [Cross Ref]

O Arah2008"The role of causal reasoning in understanding Simpson's paradox, Lord's paradox, and the suppression effect: Covariate selection in the analysis of observational studies,"Emerging Themes in Epidemiology4doi: 10.1186/1742–7622–5–5, online at <http://www.ete-online.com/content/5/1/5>.18291031

Arjas E, Parner J, authors. 2004;"Causal reasoning from longitudinal data,". Scandinavian Journal of Statistics. 31:171–187. [Cross Ref]

Avin C, Shpitser I, Pearl J, authors. 2005. "Identifiability of path-specific effects,". Proceedings of the Nineteenth International Joint Conference on Artificial Intelligence IJCAI-05 Edinburgh, UK: Morgan-Kaufmann Publishers; p. 357–363

Balke A, Pearl J, authors. 1995. "Counterfactuals and policy analysis in structural models,". Besnard P, Hanks S, editors. Uncertainty in Artificial Intelligence 11 San Francisco: Morgan Kaufmann; p. 11–18

Balke A, Pearl J, authors. 1997;"Bounds on treatment effects from studies with imperfect compliance,". Journal of the American Statistical Association. 92:1172–1176. [Cross Ref]

Baron R, Kenny D, authors. 1986;"The moderator-mediator variable distinction in social psychological research: Conceptual, strategic, and statistical considerations,". Journal of Personality and Social Psychology. 51:1173–1182. [Cross Ref] [PubMed]

Berkson J, author. 1946;"Limitations of the application of fourfold table analysis to hospital data,". Biometrics Bulletin. 2:47–53. [Cross Ref]

Bollen K, author. 1989. Structural Equations with Latent Variables. New York: John Wiley;

Brent R, Lok L, authors. 2005;"A fishing buddy for hypothesis generators,". Science. 308:523–529. [Cross Ref] [PubMed]

Cai Z, Kuroki M, authors. 2006;"Variance estimators for three 'probabilities of causation',". Risk Analysis. 25:1611–1620. [Cross Ref] [PubMed]

Cai Z, Kuroki M, authors. 2008. "On identifying total effects in the presence of latent variables and selection bias,". McAllester DA, Myllymäki P, editors. Uncertainty in Artificial Intelligence, Proceedings of the Twenty-Fourth Conference Arlington, VA: AUAI; p. 62–69

Cartwright N, author. 2007. Hunting Causes and Using Them: Approaches in Philosophy and Economics. New York, NY: Cambridge University Press;

K ChalakH White2006"An extended class of instrumental variables for the estimation of causal effects,"Technical Report Discussion Paper, UCSD, Department of Economics.

Chickering D, Pearl J, authors. 1997;"A clinician's tool for analyzing noncompliance,". Computing Science and Statistics. 29:424–431

Cole P, author. 1997;"Causality in epidemiology, health policy, and law,". Journal of Marketing Research. 27:10279–10285

Cole S, Hernán M, authors. 2002;"Fallibility in estimating direct effects,". International Journal of Epidemiology. 31:163–165. [Cross Ref] [PubMed]

Cox D, author. 1958. The Planning of Experiments. NY: John Wiley and Sons;

Cox D, Wermuth N, authors. 2004;"Causality: A statistical view,". International Statistical Review. 72:285–305

Dawid A, author. 1979;"Conditional independence in statistical theory,". Journal of the Royal Statistical Society, Series B. 41:1–31

Dawid A, author. 2000;"Causal inference without counterfactuals (with comments and rejoinder),". Journal of the American Statistical Association. 95:407–448. [Cross Ref]

Dawid A, author. 2002;"Influence diagrams for causal modelling and inference,". International Statistical Review. 70:161–189. [Cross Ref]

Duncan O, author. 1975. Introduction to Structural Equation Models. New

York: Academic Press;

Eells E, author. 1991. Probabilistic Causality. Cambridge, MA: Cambridge University Press;

Frangakis C, Rubin D, authors. 2002;"Principal stratification in causal inference,". Biometrics. 1:21–29. [Cross Ref] [PubMed]

Glymour M, Greenland S, authors. 2008. "Causal diagrams,". Rothman K, Greenland S, Lash T, editors. Modern Epidemiology Philadelphia, PA: Lippincott Williams & Wilkins; 3rd edition. p. 183–209

Goldberger A, author. 1972;"Structural equation models in the social sciences,". Econometrica: Journal of the Econometric Society. 40:979–1001

Goldberger A, author. 1973. "Structural equation models: An overview,". Goldberger A, Duncan O, editors. Structural Equation Models in the Social Sciences New York, NY: Seminar Press; p. 1–18

Greenland S, author. 1999;"Relation of probability of causation, relative risk, and doubling dose: A methodologic error that has become a social problem,". American Journal of Public Health. 89:1166–1169. [Cross Ref] [PubMed]

Greenland S, Pearl J, Robins J, authors. 1999;"Causal diagrams for epidemiologic research,". Epidemiology. 10:37–48. [Cross Ref] [PubMed]

Greenland S, Robins J, authors. 1986;"Identifiability, exchangeability, and epidemiological confounding,". International Journal of Epidemiology. 15:413–419. [Cross Ref] [PubMed]

T Haavelmo1943"The statistical implications of a system of simultaneous equations,"Econometrica11112reprinted in DF HendryMS MorganThe Foundations of Econometric AnalysisCambridge University Press477490199510.2307/1905714

Hafeman D, Schwartz S, authors. 2009;"Opening the black box: A motivation for the assessment of mediation,". International Journal of Epidemiology. 3:838–845. [Cross Ref] [PubMed]

J Halpern1998"Axiomatizing causal reasoning,"G CooperS MoralUncertainty in Artificial IntelligenceSan Francisco, CAMorgan Kaufmann202210alsoJournal of Artificial Intelligence Research12317372000

Heckman J, author. 1992. "Randomization and social policy evaluation,". Manski C, Garfinkle I, editors. Evaluations: Welfare and Training Programs Cambridge, MA: Harvard University Press; p. 201–230

Heckman J, author. 2005;"The scientific model of causality,". Sociological Methodology. 35:1–97. [Cross Ref]

Heckman J, author. 2008;"Econometric causality,". International Statistical Review. 76:1–27. [Cross Ref]

Heckman J, Navarro-Lozano S, authors. 2004;"Using matching, instrumental variables, and control functions to estimate economic choice models,". The Review of Economics and Statistics. 86:30–57. [Cross Ref]

Heckman J, Vytlacil E, authors. 2005;"Structural equations, treatment effects and econometric policy evaluation,". Econometrica. 73:669–738. [Cross Ref]

Hernán M, Cole S, authors. 2009;"Invited commentary: Causal diagrams and measurement bias,". American Journal of Epidemiology. 170:959–962. [Cross Ref] [PubMed]

Holland P, author. 1988. "Causal inference, path analysis, and recursive structural equations models,". Clogg C, editor. Sociological Methodology Washington, DC: American Sociological Association; p. 449–484. [Cross Ref]

L Hurwicz1950"Generalization of the concept of identification,"T KoopmansStatistical Inference in Dynamic Economic ModelsCowles Commission, Monograph 10New YorkWiley245257

K ImaiL KeeleT Yamamoto2008"Identification, inference, and sensitivity analysis for causal mediation effects,"Technical reportDepartment of Politics, Princeton University

Imbens G, Wooldridge J, authors. 2009;"Recent developments in the econometrics of program evaluation,". Journal of Economic Literature. 47:5–86. [Cross Ref]

Judd C, Kenny D, authors. 1981;"Process analysis: Estimating mediation in treatment evaluations,". Evaluation Review. 5:602–619. [Cross Ref]

Kaufman S, Kaufman J, MacLenose R, authors. 2009;"Analytic bounds on causal risk differences in directed acyclic graphs involving three observed binary variables,". Journal of Statistical Planning and Inference. 139:3473–3487. [Cross Ref] [PubMed]

Kiiveri H, Speed T, Carlin J, authors. 1984;"Recursive causal models,". Journal of Australian Math Society. 36:30–52. [Cross Ref]

Koopmans T, author. 1953. "Identification problems in econometric model construction,". Hood W, Koopmans T, editors. Studies in Econometric Method New York: Wiley; p. 27–48

Kuroki M, Miyakawa M, authors. 1999;"Identifiability criteria for causal effects of joint interventions,". Journal of the Royal Statistical Society. 29:105–117

Lauritzen S, author. 1996. Graphical Models. Oxford: Clarendon Press;

Lauritzen S, author. 2001. "Causal inference from graphical models,". Cox D, Kluppelberg C, editors. Complex Stochastic Systems Boca Raton, FL: Chapman and Hall/CRC Press; p. 63–107

Lindley D, author. 2002;"Seeing and doing: The concept of causation,". International Statistical Review. 70:191–214. [Cross Ref]

MacKinnon D, Fairchild A, Fritz M, authors. 2007a;"Mediation analysis,". Annual Review of Psychology. 58:593–614. [Cross Ref]

MacKinnon D, Lockwood C, Brown C, Wang W, Hoffman J, authors.

2007b;"The intermediate endpoint effect in logistic and probit regression,". Clinical Trials. 4:499–513. [Cross Ref] [PubMed]

Manski C, author. 1990;"Nonparametric bounds on treatment effects,". American Economic Review, Papers and Proceedings. 80:319–323

J Marschak1950"Statistical inference in economics,"T KoopmansStatistical Inference in Dynamic Economic ModelsNew YorkWiley150cowles Commission for Research in Economics, Monograph 10.

Meek C, Glymour C, authors. 1994;"Conditioning and intervening,". British Journal of Philosophy Science. 45:1001–1021. [Cross Ref]

Miettinen O, author. 1974;"Proportion of disease caused or prevented by a given exposure, trait, or intervention,". Journal of Epidemiology. 99:325–332

Morgan S, Winship C, authors. 2007. Counterfactuals and Causal Inference: Methods and Principles for Social Research (Analytical Methods for Social Research). New York, NY: Cambridge University Press;

Muller D, Judd C, Yzerbyt V, authors. 2005;"When moderation is mediated and mediation is moderated,". Journal of Personality and Social Psychology. 89:852–863. [Cross Ref] [PubMed]

Neyman J, author. 1923;"On the application of probability theory to agricultural experiments. Essay on principles. Section 9,". Statistical Science. 5:465–480

Pearl J, author. 1988. Probabilistic Reasoning in Intelligent Systems. San Mateo, CA: Morgan Kaufmann;

Pearl J, author. 1993a;"Comment: Graphical models, causality, and intervention,". Statistical Science. 8:266–269. [Cross Ref]

J Pearl1993b"Mediating instrumental variables,"Technical Report R-210, <http://ftp.cs.ucla.edu/pub/stat_ser/R210.pdf>, Department of Computer Science, University of California, Los Angeles.

Pearl J, author. 1995;"Causal diagrams for empirical research,". Biometrika. 82:669–710. [Cross Ref]

Pearl J, author. 1998;"Graphs, causality, and structural equation models,". Sociological Methods and Research. 27:226–284. [Cross Ref]

Pearl J, author. 2000a. Causality: Models, Reasoning, and Inference. New York: Cambridge University Press; second ed. 2009

Pearl J, author. 2000b;"Comment on A.P. Dawid's, Causal inference without counterfactuals,". Journal of the American Statistical Association. 95:428–431. [Cross Ref]

Pearl J, author. 2001. "Direct and indirect effects,". Proceedings of the Seventeenth Conference on Uncertainty in Artificial Intelligence San Francisco, CA: Morgan Kaufmann; p. 411–420

Pearl J, author. 2004. "Robustness of causal claims,". Chickering M,

Halpern J, editors. Proceedings of the Twentieth Conference Uncertainty in Artificial Intelligence Arlington, VA: AUAI Press; p. 446–453

J Pearl2009a"Causal inference in statistics: An overview,"Statistics Surveys396146<http://www.i-journals.org/ss/viewarticle.php?id=57>.10.1214/09-SS057

Pearl J, author. 2009b. Causality: Models, Reasoning, and Inference. New York: Cambridge University Press; second edition.

J Pearl2009c"Letter to the editor: Remarks on the method of propensity scores,"Statistics in Medicine2814151416<http://ftp.cs.ucla.edu/pub/stat_ser/r345-sim.pdf>.10.1002/sim.352119340847

J Pearl2009d"Myth, confusion, and science in causal analysis,"Technical Report R-348Department of Computer Science, University of California Los Angeles, CA<http://ftp.cs.ucla.edu/pub/stat_ser/r348.pdf>.

J Pearl2009e"On a class of bias-amplifying covariates that endanger effect estimates,"Technical Report R-346Department of Computer Science, University of California Los Angeles, CA<http://ftp.cs.ucla.edu/pub/stat_ser/r346.pdf>.

J Pearl2009f"On measurement bias in causal inference,"Technical Report R-357<http://ftp.cs.ucla.edu/pub/stat_ser/r357.pdf>, Department of Computer Science, University of California Los Angeles

J PearlA Paz2009"Confounding equivalence in observational

studies,"Technical Report R-343Department of Computer Science, University of California Los Angeles, CA<http://ftp.cs.ucla.edu/pub/stat_ser/r343.pdf>.

Pearl J, Robins J, authors. 1995. "Probabilistic evaluation of sequential plans from causal models with hidden variables,". Besnard P, Hanks S, editors. Uncertainty in Artificial Intelligence 11 San Francisco: Morgan Kaufmann; p. 444–453

Pearl J, Verma T, authors. 1991. "A theory of inferred causation,". Allen J, Fikes R, Sandewall E, editors. Principles of Knowledge Representation and Reasoning: Proceedings of the Second International Conference San Mateo, CA: Morgan Kaufmann; p. 441–452

Petersen M, Sinisi S, van der Laan M, authors. 2006;"Estimation of direct causal effects,". Epidemiology. 17:276–284. [Cross Ref] [PubMed]

Richard J, author. 1980;"Models with several regimes and changes in exogeneity,". Review of Economic Studies. 47:1–20. [Cross Ref]

Robertson D, author. 1997;"The common sense of cause in fact,". Texas Law Review. 75:1765–1800

Robins J, author. 1986;"A new approach to causal inference in mortality studies with a sustained exposure period – applications to control of the healthy workers survivor effect,". Mathematical Modeling. 7:1393–1512. [Cross Ref]

Robins J, author. 1987;"A graphical approach to the identification and estimation of causal parameters in mortality studies with sustained exposure periods,". Journal of Chronic Diseases. 40:139S–161S. [Cross Ref] [PubMed]

Robins J, author. 1989. "The analysis of randomized and non-randomized aids treatment trials using a new approach to causal inference in longitudinal studies,". Sechrest L, Freeman H, Mulley A, editors. Health Service Research Methodology: A Focus on AIDS Washington, DC: NCHSR, U.S. Public Health Service; p. 113–159

Robins J, author. 1999. "Testing and estimation of directed effects by reparameterizing directed acyclic with structural nested models,". Glymour C, Cooper G, editors. Computation, Causation, and Discovery Cambridge, MA: AAAI/MIT Press; p. 349–405

Robins J, author. 2001;"Data, design, and background knowledge in etiologic inference,". Epidemiology. 12:313–320. [Cross Ref] [PubMed]

Robins J, Greenland S, authors. 1989a;"Estimability and estimation of excess and etiologic fractions,". Statistics in Medicine. 8:845–859. [Cross Ref] [PubMed]

Robins J, Greenland S, authors. 1989b;"The probability of causation under a stochastic model for individual risk,". Biometrics. 45:1125–1138. [Cross Ref] [PubMed]

Robins J, Greenland S, authors. 1992;"Identifiability and exchangeability for direct and indirect effects,". Epidemiology. 3:143–155. [Cross Ref] [PubMed]

Rosenbaum P, author. 2002. Observational Studies. New York: Springer-

Verlag; second edition.

Rosenbaum P, Rubin D, authors. 1983;"The central role of propensity score in observational studies for causal effects,". Biometrika. 70:41–55. [Cross Ref]

Rothman K, author. 1976;"Causes,". American Journal of Epidemiology. 104:587–592. [PubMed]

Rubin D, author. 1974;"Estimating causal effects of treatments in randomized and non-randomized studies,". Journal of Educational Psychology. 66:688–701. [Cross Ref]

Rubin D, author. 2004;"Direct and indirect causal effects via potential outcomes,". Scandinavian Journal of Statistics. 31:161–170. [Cross Ref]

Rubin D, author. 2005;"Causal inference using potential outcomes: Design, modeling, decisions,". Journal of the American Statistical Association. 100:322–331. [Cross Ref]

Rubin D, author. 2007;"The design *versus* the analysis of observational studies for causal effects: Parallels with the design of randomized trials,". Statistics in Medicine. 26:20–36. [Cross Ref] [PubMed]

Rubin D, author. 2009;"Author's reply: Should observational studies be designed to allow lack of balance in covariate distributions across treatment group?". Statistics in Medicine. 28:1420–1423. [Cross Ref]

Shpitser I, Pearl J, authors. 2006. "Identification of conditional interventional distributions,". Dechter R, Richardson T, editors. Proceedings of the Twenty-Second Conference on Uncertainty in Artificial Intelligence Corvallis, OR: AUAI Press; p. 437–444

Shpitser I, Pearl J, authors. 2008. "Dormant independence,". Proceedings of the Twenty-Third Conference on Artificial Intelligence Menlo Park, CA: AAAI Press; p. 1081–1087

Shpitser I, Pearl J, authors. 2009. "Effects of treatment on the treated: Identification and generalization,". Proceedings of the Twenty-Fifth Conference on Uncertainty in Artificial Intelligence Montreal, Quebec: AUAI Press;

I Shrier2009"Letter to the editor: Propensity scores,"Statistics in Medicine2813171318see also Pearl 2009<http://ftp.cs.ucla.edu/pub/stat_ser/r348.pdf>.10.1002/sim.355419288531

Shrout P, Bolger N, authors. 2002;"Mediation in experimental and nonexperimental studies: New procedures and recommendations,". Psychological Methods. 7:422–445. [Cross Ref] [PubMed]

Simon H, author. 1953. "Causal ordering and identifiability,". Hood WC, Koopmans T, editors. Studies in Econometric Method New York, NY: Wiley and Sons, Inc; p. 49–74

Simon H, Rescher N, authors. 1966;"Cause and counterfactual,". Philosophy and Science. 33:323–340. [Cross Ref]

Sobel M, author. 1998;"Causal inference in statistical models of the process of socioeconomic achievement,". Sociological Methods & Research. 27:318–348. [Cross Ref]

Sobel M, author. 2008;"Identification of causal parameters in randomized studies with mediating variables,". Journal of Educational and Behavioral Statistics. 33:230–231. [Cross Ref]

Spirtes P, Glymour C, Scheines R, authors. 1993. Causation, Prediction, and Search. New York: Springer-Verlag;

Spirtes P, Glymour C, Scheines R, authors. 2000. Causation, Prediction, and Search. Cambridge, MA: MIT Press; 2nd edition.

Stock J, Watson M, authors. 2003. Introduction to Econometrics. New York: Addison Wesley;

Strotz R, Wold H, authors. 1960;"Recursive versus nonrecursive systems: An attempt at synthesis,". Econometrica. 28:417–427. [Cross Ref]

Suppes P, author. 1970. A Probabilistic Theory of Causality. Amsterdam: North-Holland Publishing Co;

J TianA PazJ Pearl1998"Finding minimal separating sets,"Technical Report R-254, University of California Los Angeles, CA

Tian J, Pearl J, authors. 2000;"Probabilities of causation: Bounds and identification,". Annals of Mathematics and Artificial Intelligence. 28:287–313.

[Cross Ref]

Tian J, Pearl J, authors. 2002. "A general identification condition for causal effects,". Proceedings of the Eighteenth National Conference on Artificial Intelligence Menlo Park, CA: AAAI Press/The MIT Press; p. 567–573

VanderWeele T, author. 2009;"Marginal structural models for the estimation of direct and indirect effects,". Epidemiology. 20:18–26. [Cross Ref] [PubMed]

VanderWeele T, Robins J, authors. 2007;"Four types of effect modification: A classification based on directed acyclic graphs,". Epidemiology. 18:561–568. [Cross Ref] [PubMed]

T VermaJ Pearl1990"Equivalence and synthesis of causal models,"Proceedings of the Sixth Conference on Uncertainty in Artificial IntelligenceCambridge, MA220227also in P BonissoneM HenrionLN KanalJF LemmerUncertainty in Artificial Intelligence 6Elsevier Science Publishers, B.V.2552681991

N Wermuth1992"On block-recursive regression equations,"Brazilian Journal of Probability and Statistics(with discussion)6156

Wermuth N, Cox D, authors. 1993;"Linear dependencies represented by chain graphs,". Statistical Science. 8:204–218. [Cross Ref]

Whittaker J, author. 1990. Graphical Models in Applied Multivariate Statistics. Chichester, England: John Wiley;

Wilkinson L; the Task Force on Statistical Inference and APA Board of Scientific Affairs, authors. 1999;"Statistical methods in psychology journals: Guidelines and explanations,". American Psychologist. 54:594–604. [Cross Ref]

Woodward J, author. 2003. Making Things Happen. New York, NY: Oxford University Press;

Wooldridge J, author. 2002. Econometric Analysis of Cross Section and Panel Data. Cambridge and London: MIT Press;

J Wooldridge2009"Should instrumental variables be used as matching variables?"Technical Report <https://www.msu.edu/~ec/faculty/wooldridge/current%20research/treat1r6.pdf>Michigan State University, MI

Wright S, author. 1921;"Correlation and causation,". Journal of Agricultural Research. 20:557–585